WITHDRAWN

EXPLORING WITH LASERS

EXPLORING WITH LASERS

Brent Filson

Illustrations by
Brigita Fuhrmann

Julian Messner   New York

Copyright © 1984 by Brent Filson
All rights reserved including the right of
reproduction in whole or in part in any form.
Published by Julian Messner, A Division of Simon & Schuster, Inc.
Simon & Schuster Building,
1230 Avenue of the Americas,
New York, New York 10020.
JULIAN MESSNER and colophon are trademarks of
Simon & Schuster, Inc.

Manufactured in the United States of America

10   9   8   7   6   5   4

Design by A Good Thing, Inc.

Library of Congress Cataloging in Publication Data

Filson, Brent.
   Exploring with lasers.

   Bibliography: p.
   Includes index.
   Summary: Describes the characteristics of lasers,
how they are made, how they work, and how they are used.
   1. Lasers—Juvenile literature. [1. Lasers]
I. Fuhrmann, Brigita, ill. II. Title.
TA1682.F55   1984   621.36′6   84-14731
ISBN 0-671-50573-4

For Marguerite and Floyd

CONTENTS

1 | Lasers • 9 •
2 | What is Light? • 11 •
3 | Laser Light • 19 •
4 | Constructing a Laser • 24 •
5 | Types of Lasers • 28 •
6 | Medical Uses of Lasers • 33 •
7 | Military Uses of Lasers • 39 •
8 | Communications • 45 •
9 | Lasers in Industry • 49 •
10 | Measurement by Lasers • 56 •
11 | Holograms and Lasers • 61 •
12 | A Garden of Laser Delights • 69 •
13 | Lasers in Art and Entertainment • 73 •
14 | Laser Safety • 75 •
15 | The Future of Lasers • 78 •

Glossary • 86 •
Bibliography • 93 •
Index • 94 •

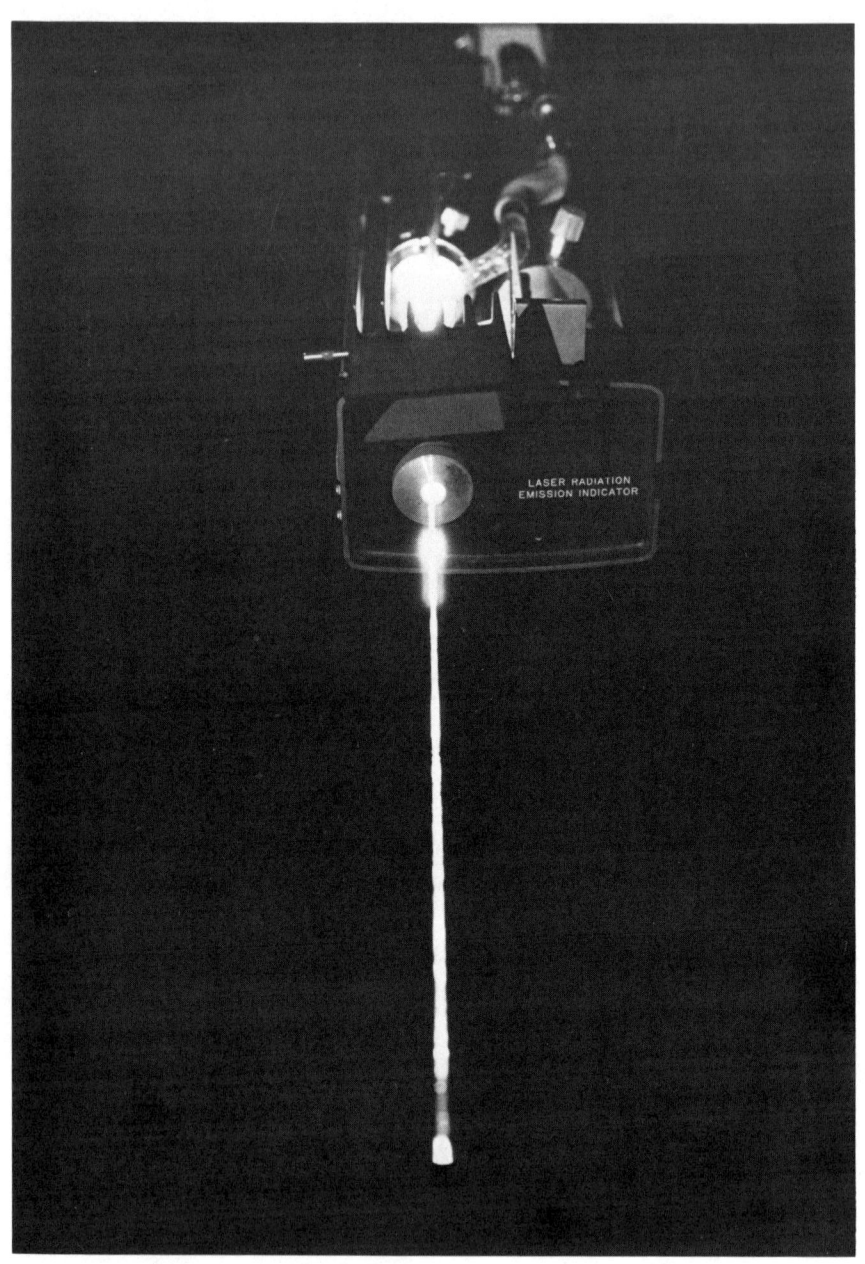

A laser beam emerges from an argon ion laser. The laser's housing has been removed for the purposes of the photo, exposing the plasma tube. (Jason Sapan/ Holographic Studios, NY)

• 1 •

LASERS

A laser is a special kind of light. Think of it as a rod of light. It doesn't spread out as normal light does. It goes in one direction. And it can be very intense. In addition, a beam of laser light is composed of only a single color.

A laser can do many things. It can be used to perform delicate eye surgery. It can also trigger a thermonuclear explosion. It can carry more than one hundred thousand times more information than a single telephone wire. It can create photographs in three dimensions. It can produce twenty thousand lines of print in a minute. It can drill through something as hard as a diamond or as soft as a

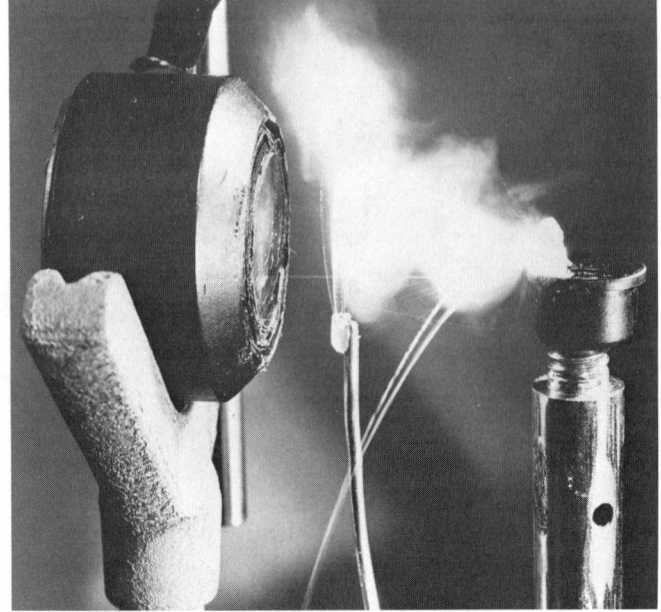

A laser burns a hole in a diamond. The laser is out of the picture at left. The beam's intensity is increased by focusing it through an optical lens. (General Electric Research and Development Center)

baby-bottle nipple. It can measure time in trillionths of a second or detect the vibration of an electron. And some scientists say that there may come a day when the laser will help us communicate with distant star systems, propel rockets and airplanes faster and more cheaply than ever before, and provide the world with unlimited energy.

Yet just a little more than two decades ago, nobody had ever seen a laser beam. Lasers did not exist on earth. The fact that they could exist was only a theory.

In this book, you will learn about this exciting new tool of light; how it is made, how it works, and how it is used.

• 2 •
WHAT IS LIGHT?

In order to understand what a laser is, it is first necessary to examine what light itself is. Light is all around us. It comes from as far away as the sun and stars. It comes off the page of this book. Light fills our most important sense, our vision, every waking moment. But the question—what is light?—has puzzled people for thousands of years. Over the centuries, the answer has been hotly debated. The answers to the question, What is light?, could almost be a history of physics.

Ancient peoples around the world had many ideas about the nature of light. For instance, about two thousand years ago, some Greeks and Hindus said that light was something that poured out of the eyes. These people reasoned that the eyes "felt" objects in front of them. Other Greeks said that light was made up of tiny particles that flowed into the eyes.

One of the great Greek thinkers, Plato, had a more complex view. He said that light was made up of particles that streamed from the eyes and of rays of sunlight as well. (See Fig. 1) He said that sun rays then combined with these particles and entered the eyes and triggered vision. For over a thousand years, people were convinced that Plato's idea of light was correct.

But in the eleventh century A.D., a famous scientist said that Plato was wrong. That scientist was an Arabian named Al-Hazen. Al-Hazen studied the eye and detailed how it promoted vision. After much study, Al-Hazen said that particles do not stream out of the eye at all. Objects are seen, he declared, because light shines off them and then into the eye.

But that still left the question: What exactly is light? Some five hundred years later, in the late 1600s, two scientists became convinced that they

Fig. 1 Plato's view of light.

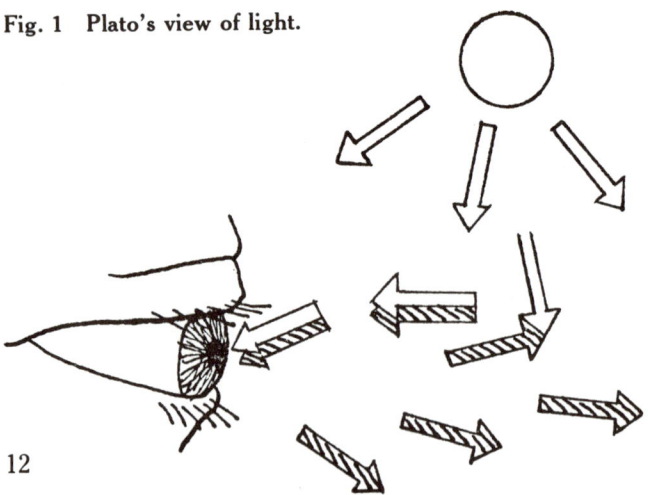

had the answer. They each had different answers and each one fueled a debate that would go on for several centuries. In fact, in many ways that debate has not died down yet.

One of these men was the great Dutch physicist, Christian Huygens. Huygens did important work with microscopes and telescopes. He knew that when light passed through the lenses of these instruments, it spread out and bent. Since waves are things that spread out and bend, Huygens reasoned that light must be composed of waves. This was called the wave theory of light.

At about the same time, another scientist came up with an opposite view. His name was Isaac Newton. Newton is probably best known for formulating the law of gravity. But he also made important discoveries concerning light. For instance, he discovered that white or normal light is in fact a mixture of different-colored light rays. He also maintained that Huygens was wrong about the nature of light. Newton said that since light travels in straight lines (or rays), it must be composed not of waves, but of tiny chunks or particles. This was called the particle theory.

These opposing theories of light—the wave theory and the particle theory—caused an uproar among scientists. Some sided with Huygens. Others sided with Newton. The debate raged for over a hundred years. Then in 1803, an Englishman named Thomas Young claimed to settle the issue once and for all. He set up a simple but clever experiment. (See Fig. 2)

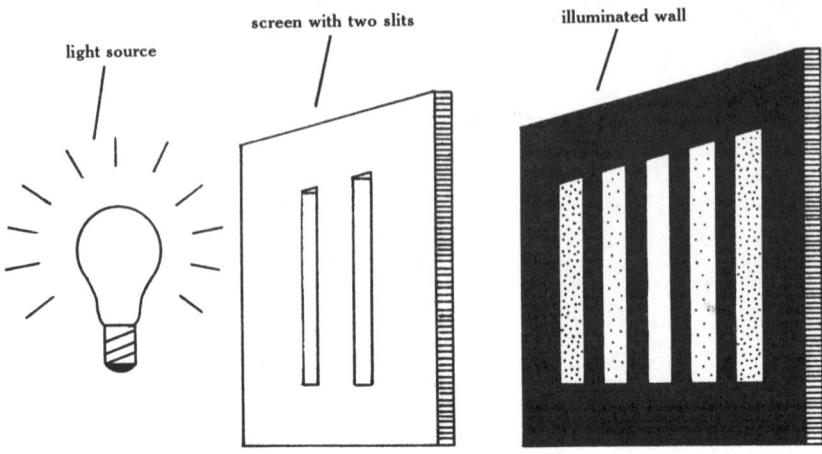

Fig. 2 Thomas Young's experiment.

By directing sunlight through two slits in a screen and onto a wall, he showed an interesting phenomenon. You would think that if the sunlight went through two slits, two bands of light would appear on the wall. But this isn't what happens when that experiment is performed. What happens instead is that five bands of light appear on the wall. Some bands are dark. Others are bright. This means that the two slices of sunlight did not pass in a straight line to the wall. Instead, they actually got in the way of each other after they passed through the slits. They *interfered* with each other. (We'll examine this important aspect of light, *interference,* when we see how lasers help create three-dimensional photographs). Thus the light acted much like waves do.

If you drop two stones, side by side, into water, two sets of waves spread out and run against each other. Where the crests of the waves meet, larger waves are formed. Where the troughs of the waves of one set meet the crests of the waves of another, the water becomes somewhat calm. Waves interfere with each other. Thus the result of Young's experiment is clear. Light is composed of waves.

The issue finally seemed to be settled. But it wasn't really settled at all. For there was one question that nagged at the scientists. At first, it seemed like an easy question to answer. Most of the scientists thought that once it was answered, everything about light would be known. But as it turned out, that question would trigger one of the greatest wild-goose chases in the history of science.

The question was this: If light was composed of waves, then what was the substance the waves vibrated in? After all, waves must vibrate in something. Sound waves, for instance, need air to vibrate in. Sound can be heard under water because there it has a medium in which to vibrate. In space, there is no air, and thus there is no sound. But light did not need air to exist. Light moves through outer space. It moves through vacuum jars. What was it that light waves passed through?

Fig. 3

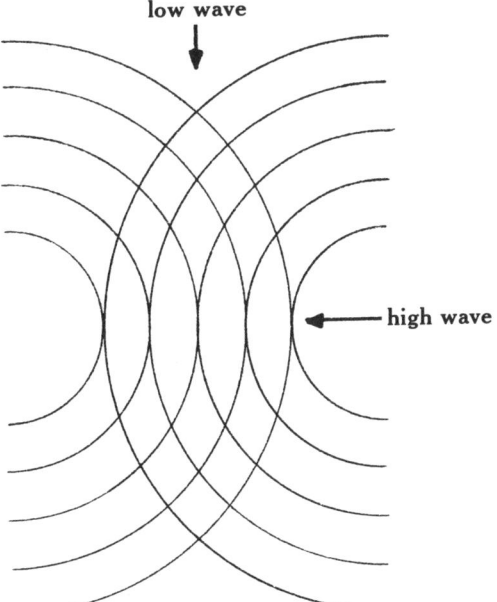

Scientists made up an answer. They figured that there must be some substance that fills space and allows light waves to vibrate. They called that substance *ether*. They said that it was odorless, colorless, invisible, and could not be felt. They reasoned that since light traveled throughout our world and throughout space, ether was everywhere. It seemed like a foolproof answer. The only thing needed was to prove ether existed.

During the last century, many experiments were conducted and mathematical formulas created to prove the existence of ether. Finally, in 1905, one man, working by himself, came up with the answer. That man was Albert Einstein. And his answer was very simple. It was an answer that shook the very foundations of science and helped bring about a whole new view of the universe. Einstein's answer was that ether didn't exist.

Albert Einstein was one of the greatest scientists of all time. He developed important ideas about the way we think of motion, space, and time. One of his most important ideas deals with light. In 1905, he wrote a paper about light that was to win him the Nobel Prize. Drawing on the work of such important scientists as James Maxwell, Max Planck, and Philipp Lenard, Einstein said that a beam of light is actually like a stream of bullets. The bullets are tiny particles of energy. He called the particles *photons*. Einstein's theory was soon proved to be true. Light was actually made up of photons. Light didn't need something to vibrate in. Scientists couldn't prove that ether existed simply because it did not exist.

Albert Einstein (National Archives Courtesy AIP Niels Bohr Library)

Thus in nearly 250 years, the theories about light had come full circle back to the ideas of Newton.

But wait a moment. Remember Young's experiment in 1803? That experiment proved that light was made up of waves. Einstein proved that it is made up of particles. Who was right?

Einstein had an answer for that too. He said that just because he proved light is made up of tiny particles, that does not mean light is not a wave. Logical or not, light can act as either a wave or a particle—depending on the experiment.

One way to understand this is to picture a hurdler. The hurdler herself is a chunk of matter. But as she jumps over the hurdles, her action can be traced with the crests and troughs of a wave. So she can have both particle and wave aspects, just as light does.

This answer may not satisfy some people. Many scientists are not content that we can understand light in two opposing ways. They say that this describes what light *does,* not what light *is.* So, in this sense, the debate still goes on.

◆ 3 ◆
LASER LIGHT

In 1917, Albert Einstein developed another idea concerning light. He wrote that a special kind of light could be created under the right conditions. It was a light nobody had seen before. The light, he said, would be a single color. It would not scatter the way normal light does. And it would be very intense.

This kind of light would later be called laser light. How could Einstein imagine a kind of light that nobody had ever seen before? As he did so many times when he created new ideas, Einstein combined known facts with creative guesses.

The facts Einstein had to work with were these. Although atoms are much too small to be seen, even with the most powerful microscope, they are nonetheless the building blocks of the universe. All matter is made up of atoms. Every atom consists of a nucleus around which one or more electrons orbit. When electrons are stimulated, with heat, for example, they give off bursts of energy.

Up until 1913, scientists thought that electrons tumbled randomly around atoms. But in that year, a Danish physicist named Niels Bohr made an important discovery. He said that electrons do not move randomly. Instead, they circle the atoms in set orbits. These orbits can be altered only when energy is exchanged.

Take the simplest atom, for instance, a hydrogen atom. A hydrogen atom consists of a nucleus and one orbiting electron. (See Fig. 4) Normally, the electron will orbit close to the nucleus. This orbit is called the *ground state.* But when the atom is stimulated, the electron absorbs the energy and jumps to a higher orbit. This process is called *absorption.*

Fig. 4 Hydrogen atom.

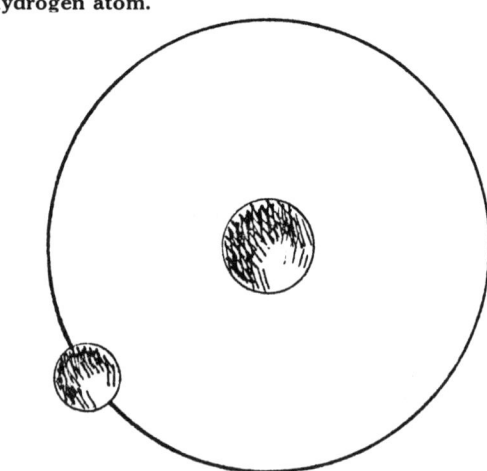

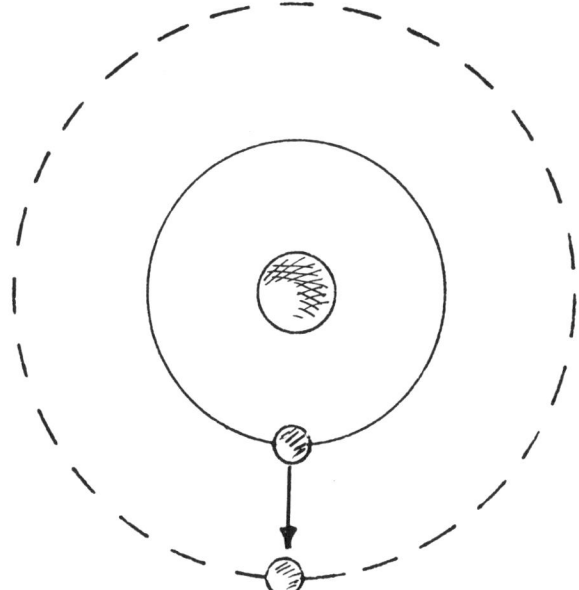

Fig. 5 Absorption.

(See Fig. 5) The greater the stimulation, the higher the electron jumps. Within a fraction of a second, the excited electron jumps back to its ground-state orbit. During this jump, a remarkable thing happens. The electron emits a burst of energy. That burst of energy takes the form of light. This is the photon that Einstein described.

For example, when a flame touches metal, the atoms that make up the metal are stimulated. Their electrons jump to higher orbits. In jumping back, they emit photons, and thus the metal glows. This process is called *spontaneous emission*. All normal light is created through spontaneous emission.

These facts were known in 1917 when Einstein wrote his paper. But he took these facts and made a leap of creative thought. He said that there might be another kind of emission of photons. He called it a *stimulated emission.* He theorized that if an excited electron was hit by a photon of a specific energy level, two important things would happen. The photon would not be absorbed. Instead, it would go on its own way. (See Fig. 6B) And the electron would not jump to a higher orbit. Instead, it would drop back to a lower orbit. In doing so, it would emit another photon. As a result, two photons would fly away from the same atom. (See Fig. 6C) Both photons would be remarkably similar. They would have the same wave length. They would also have the same phase. (In other words, their wavelengths would be in step with each other.) And finally, they would move in the same direction.

Without having seen it, Albert Einstein had described laser light.

But why had nobody seen laser light? Why did it take another forty-five years before a device could be made to create it?

The answer lies in the fact that atoms always tend to be at their ground state. When they are stimulated, they remain at an excited state for only a fraction of a second. The chances of a photon with the same energy level striking an excited electron are very small. Thus we see no laser light from natural causes. (There is a natural laser pulsing in the atmosphere of Mars. We'll talk about that in chapter 15.) In fact, in the years following the publication of Einstein's paper on stimulated emissions, many scientists thought it would be impossible to create the special conditions to make a laser.

But in 1960, those conditions were created. The first laser on earth was fired.

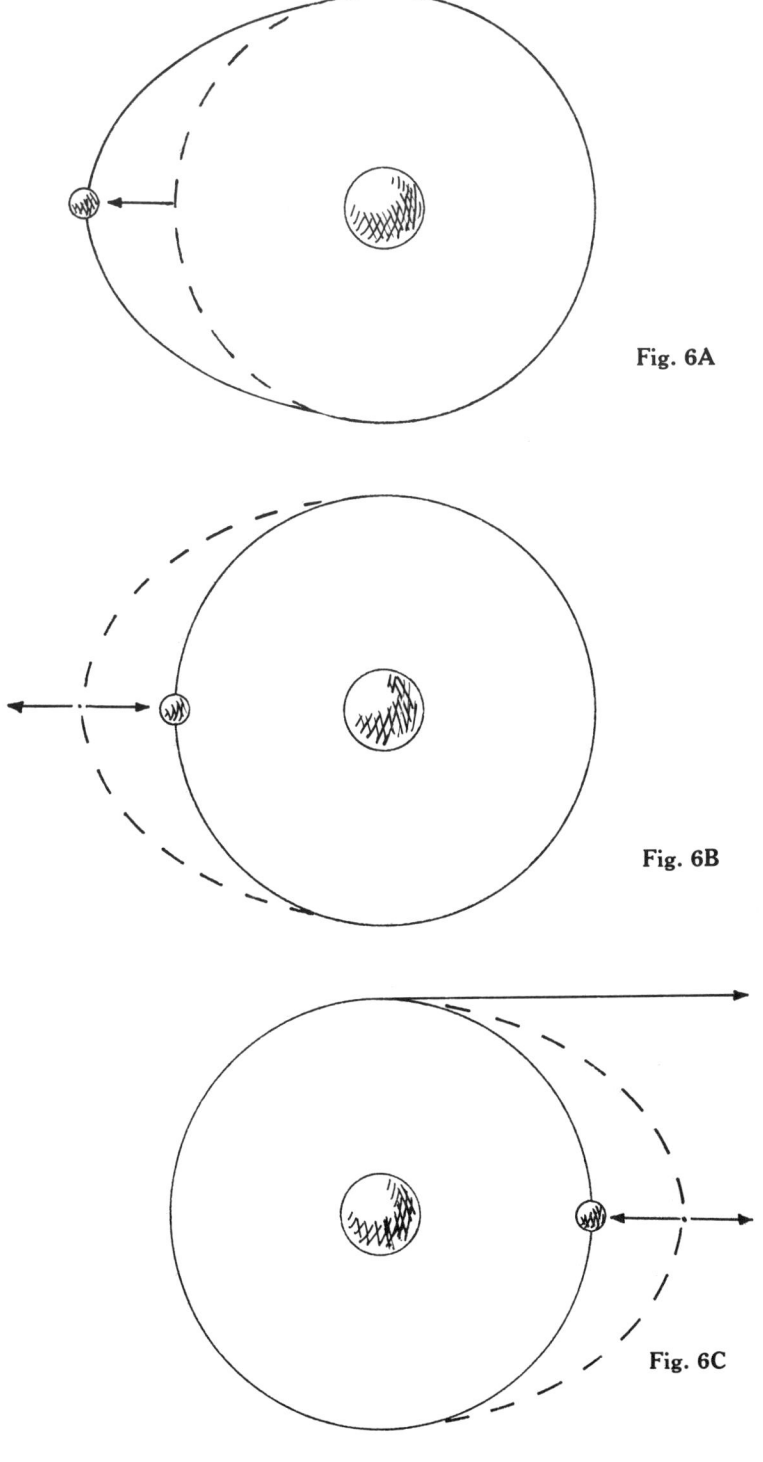

Fig. 6A

Fig. 6B

Fig. 6C

◆ 4 ◆
CONSTRUCTING A LASER

So far, we have examined the nature of light. We have seen that there are two kinds of light: normal light and laser light. We have looked at how the actions of electrons and photons create the differences in these two kinds of light. Now we'll examine how to make a laser.

In order to do this, let's first look at the word laser. Laser is an acronym. An acronym is a word formed from the first letters in a set phrase. L-A-S-E-R means: Light Amplification by Stimulated Emission of Radiation. Having read the first three chapters, you should have a good idea of what light amplification by stimulated emission means. What about the R—radiation? Let's examine that.

As was mentioned earlier, light is a form of energy. That energy is a phenomenon that involves electric currents and magnetic fields. It is called *electromagnetic* energy. Electromagnetic energy exists in various lengths of waves. The movement of these waves is called electromagnetic radiation. Radio waves are one form of electromagnetic radiation. X-rays are another. Still another are gamma rays. Finally, there are light waves. So when we speak of radiation in terms of lasers, we mean light-wave radiation.

The difference between each kind of electromagnetic wave is found in the length of the wave or in its frequency. Frequency means how many wave crests pass a given point each second. For instance, a normal AM radio wave is several miles (or kilometers) long. Because it is long, its frequency is much lower than, say, gamma rays. Gamma rays are so short that one hundred billion of them take up only one meter. Individual light waves are too short to be seen with the naked eye. But their total effect makes up the visible part of the electromagnetic spectrum. In other words, light waves are the only electromagnetic radiation we can see.

In order to trigger a beam of laser light, a special atomic condition must be created. Remember that when atoms are in their ground state, their electrons absorb photons that stimulate them. But when atoms are in a special, excited state (with their electrons whirling in distant orbits), then spontaneous emission may take place. This special, excited state is called a *population inversion*. It sounds complicated, but it is really quite simple. A population inversion means that more atoms are in a specific excited state than in their normal or ground state.

The first population inversion was created in 1954 by three American scientists. It was called a maser. Instead of light waves, microwaves were utilized. Because microwaves are invisible, the maser beam could not be seen. But in 1960, the first stimulated emission of light waves was created. It was done by a California physicist, Theodore Maiman. Maiman used a crystal of ruby, a rod only a few inches long. Around the rod he wound coils of a flash lamp, which emits a brief, very powerful burst of light when it is triggered, like the flash cubes on a camera. He coated both ends of the rod with silver to reflect light. When Maiman switched on the power and triggered the flash lamp, a beam of laser light shot from the cylinder. The beam lasted only a few millionths of a second. But it was the first laser.

This is what happened inside the crystal. (See Fig. 7) When the flash lamp was fired, a burst of intense light shot through the ruby. Atoms inside the ruby were stimulated, and electrons were excited into higher orbits. They remained in this state for a very brief time, only about one ten-millionth of a second. If normal light was to be created, the atoms would drop back to ground state, with each electron emitting a photon. But the intense light from the flash lamp had created the conditions for laser action.

The electrons dropped back to a lower energy level where *lasing* can take place. They "lingered" at this level for a thousandth of a second. This is a relatively long time in terms of photon emission. It was during this time that the electrons were stimu-

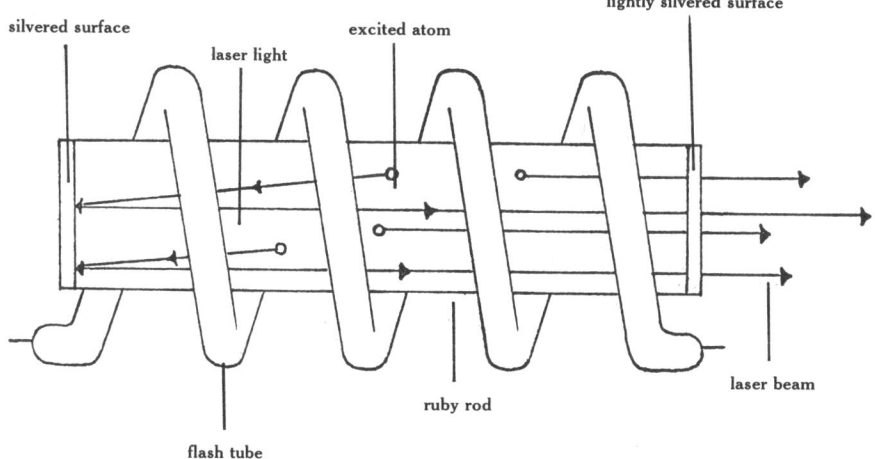

Fig. 7 Ruby laser.

lated in a special way. Instead of absorbing photons, they emitted them. Many of these photons passed out of the ruby rod in the form of heat. But many traveled along the rod in the form of laser light.

Initially, this light was rather weak. But the mirrored ends of the rod caused the light to bounce back and forth. As it did, more and more laser light (and thus energy) built up inside the rod. An explosion could have resulted if the light hadn't been released. Maiman provided an opening for the light to go out by a simple, ingenious device. He coated one of the two mirrors with a reduced mount of silver. Thus, when a strong pulse of laser light built up in the rod, it shot through the partially silvered mirror—and a new chapter in the history of science was opened.

• 5 •
TYPES OF LASERS

Since 1960, many different kinds of lasers have been developed. There are lasers that can handle forty thousand telephone calls through a single, optic fiber. There are lasers that can measure the distance between the moon and the earth within a fraction of a millimeter. There are lasers that "read" and that make three-dimensional photographs. There are lasers that destroy and lasers that heal. Scientists are working on lasers that may sometime in the future help generate unlimited energy.

Basically, there are three different kinds of lasers: solid, liquid, and gas. Here are some of the most important types.

OPTICALLY PUMPED, SOLID LASERS

Theodore Maiman's laser is called an optically pumped solid laser. This means that the atoms of a solid material, usually crystal or glass, are excited or pumped by bursts of light. In all lasers of this type, the solid material consists of a rod with mirrors at the ends to make the light bounce back and forth. Since crystal and glass can easily be damaged by overheating, this laser usually does not run continuously. Instead, it fires pulses of laser light. In between each pulse, the rod cools a little to prevent its being damaged.

High-powered pulses can be made by these types of lasers. One way of increasing the power of the pulse is by using a Q switch. A Q switch keeps the laser light from escaping until a tremendous amount of power is built up. Then the switch—in some cases, a spinning mirror—is suddenly opened up. Lasers with a Q switch can deliver a several-billion-watt laser pulse lasting a few billionths of a second.

Even more power can be created by "mode-locking." A laser pulse that is made stronger by mode-locking lasts only a few trillionths of a second, but it can produce more power in that brief moment than all the world's electric power stations combined.

There are, however, optically pumped crystal lasers that, when water-cooled and running on much less power, can produce a continuous beam. These lasers can be used as range finders and, in more powerful types, as drills.

LIQUID LASERS

In these types of lasers, a special liquid dye takes the place of the solid rod. The dye is put inside a glass tube. Like the solid lasers, pumping is accomplished with bursts of light. But unlike the solid lasers, liquid ones can run for longer periods with a continuous beam of light. They don't break down as frequently as solid types of lasers do. Furthermore, because different colored dyes can be injected into the tube, light waves of different frequencies can be created. In this way, the laser can be tuned to a variety of light frequencies.

Fig. 8 Liquid laser.

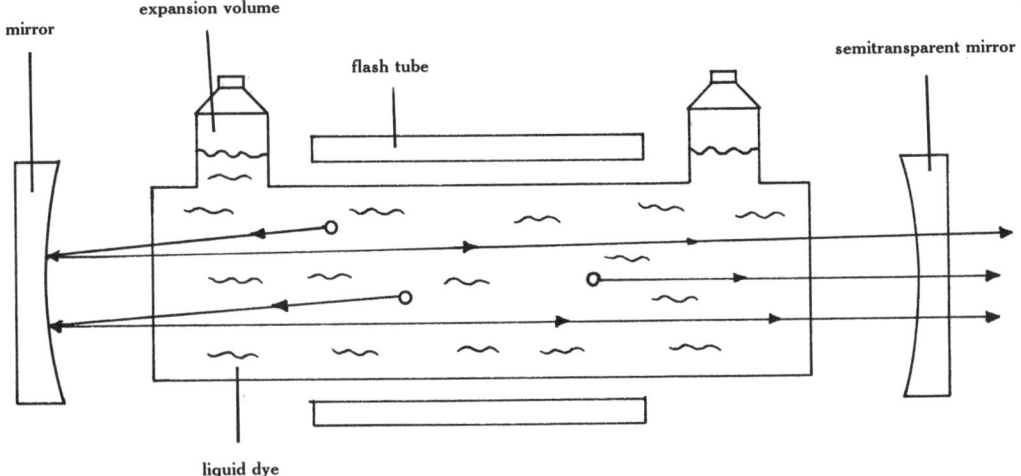

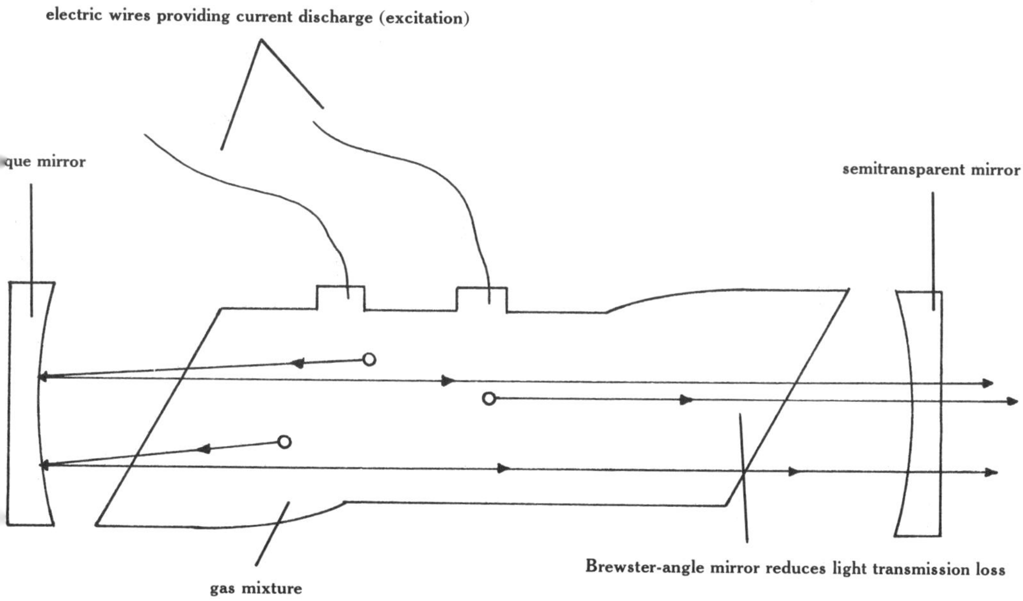

electric wires providing current discharge (excitation)

que mirror

semitransparent mirror

gas mixture

Brewster-angle mirror reduces light transmission loss

Fig. 9 Gas laser.

GAS LASERS

Gas lasers have the most coherent or narrow beam of any type of laser. Therefore, they are used in measuring and communications. Gas or gas mixtures are put in a tube (some as long as 30 feet [9 meters]) and are then stimulated by one of several power sources: electrical, chemical, electron beams, or ultraviolet rays. Lower-power helium neon gas lasers, which produce a red beam, are being used in supermarket checkout counters to read product labels. Stronger gas lasers—such as carbon dioxide—can do welding, drilling, and cutting, and are also employed in surgery as unique scalpels. Continuous, high-power beaming can be created with a gas laser by forcing the gases through the device at supersonic speeds.

This tiny semiconductor laser is shown amid single grains of common household salt. It can be "tuned" from one frequency to another. (AT&T Bell Laboratories)

Other types of lasers include *chemical* lasers, which rely on chemical reactions to carry out pumping. These can be used in remote areas, such as space, where there are no electrical power sources. *Semiconductor* lasers, some of which are smaller than a grain of salt, are inexpensive and useful in electronics.

◆ 6 ◆
MEDICAL USES OF LASERS

In 1963, a researcher gave a vivid demonstration of a laser in action. He aimed a laser at two balloons. One balloon was clear, the other was blue. The blue balloon was inflated inside the clear one. To the surprise of the reporters, the beam went through the clear, outer balloon and caused only the blue, inner balloon to burst.

It wasn't magic, but scientific fact. The reason was obvious, if you knew something about lasers. The laser was an optically pumped ruby laser. The color blue absorbed the energy of the laser's beam. When the beam struck the clear, outer balloon, it passed harmlessly through, as a normal light ray would. But when it struck the inner balloon, it

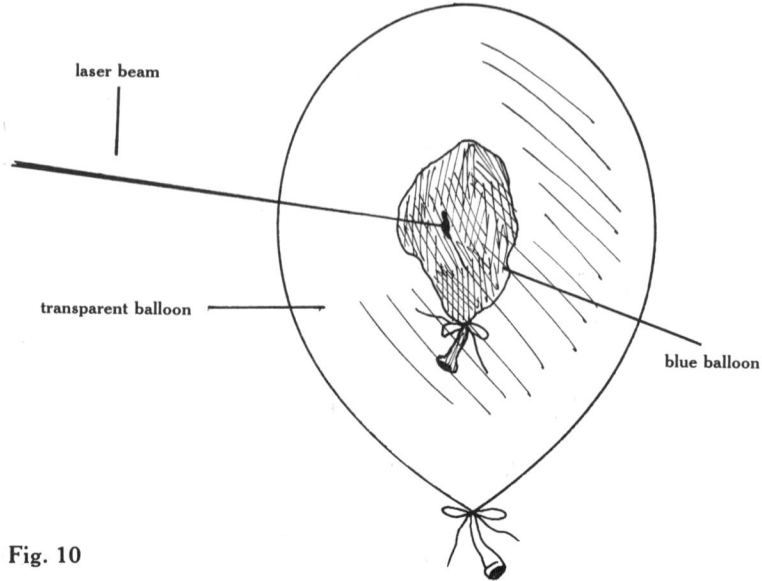

Fig. 10

punched a hole in its skin. This was because the inner balloon's blue color absorbed the energy of the ruby laser.

The demonstration was more than just a publicity stunt. It was evidence of one of the most important medical uses of lasers: their ability to repair damaged eyes.

In a rough way, the structure of the human eye is like the inner and outer balloons of the demonstration. The outer, clear portion is the cornea. The inner part is the retina. The eye sees by images of the outer world streaming through the clear cornea and being projected onto the retina. The retina, composed of light-sensitive rods and cones, acts like a movie screen onto which light is projected. Images then travel along the nerves from the retina to the brain. The retina is sensitive tissue. Sometimes a piece of it tears loose. The result is the same as if a movie screen tears apart: The image that is

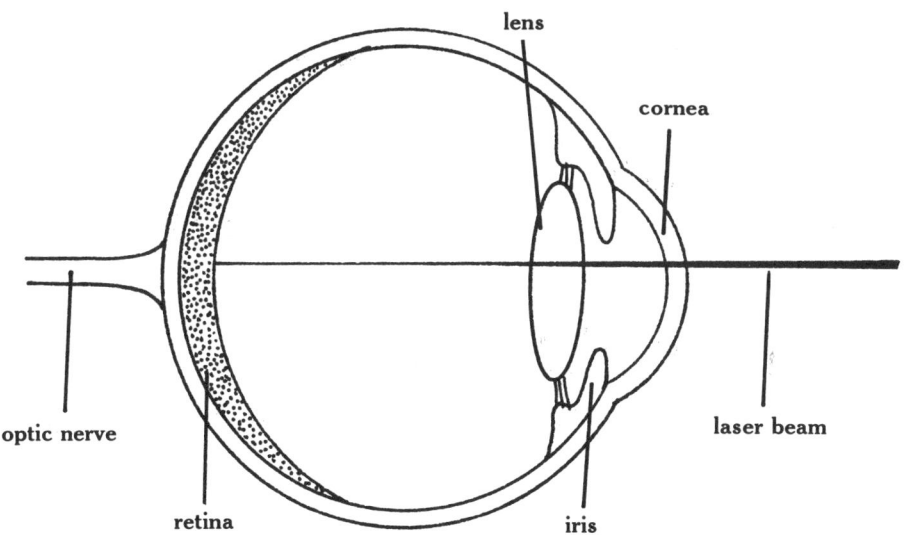

Fig. 11 Laser beam reaching retina through cornea and lens.

thrown on it will be distorted. People with torn retinas can go blind.

Repairing a torn retina used to be difficult and time-consuming. Today a laser can make the repair almost routine. Just like the laser beam that passed harmlessly through the clear outer balloon and then broke the blue inner one, a laser can pass right through the cornea. It then welds the dark-colored retina back into place. Before lasers were used in medicine, repairing a damaged retina required a several-week stay in the hospital. Today, patients can go home within a few days after laser repair of their retinas.

Lasers have many other uses in medicine. Laser beams have replaced scalpels in some types of operations. Scalpels open tissue by cutting. Lasers, on the other hand, can vaporize tissue. It's done this way. Most of the time, a carbon dioxide laser is used. Just as the color blue absorbs the energy of a

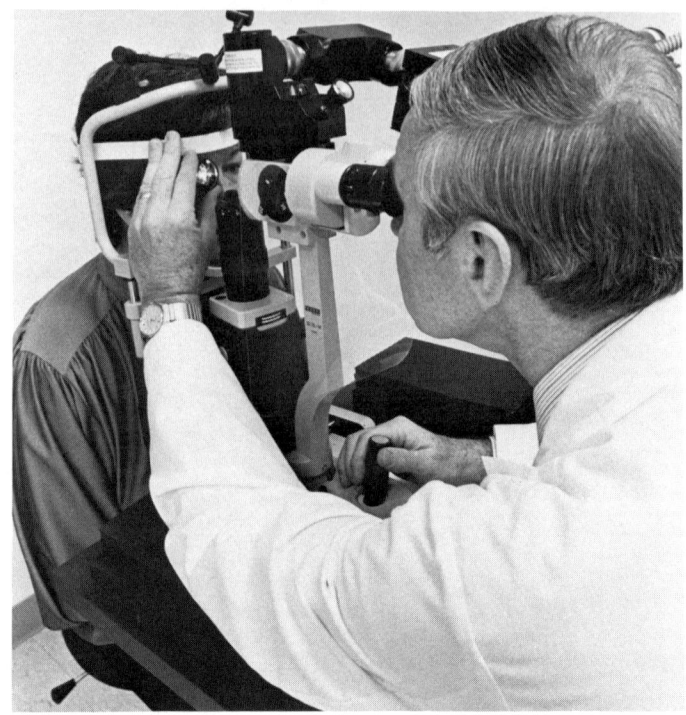

A surgeon repairs a torn retina by firing a beam of "healing" light from an argon laser. Throughout the surgery, the patient remains awake and alert. (Lahey Clinic Medical Center)

ruby laser, so water absorbs the energy of a carbon dioxide laser.

The cells of animal tissues are composed of between 75 and 90 percent water. When the laser beam strikes cells, the water turns to steam. The cells disappear in vapor, reducing the chance of infection. Of course, the beam is highly focused. The surgeon can make a cut deep or shallow. Since human cells do not conduct heat, cells not touched by the beam are undamaged. In addition, since the beam seals off tiny blood vessels, laser surgery is relatively bloodless. When an optic fiber is used to direct the beam, a laser can sometimes be used to perform surgery on hard-to-get-at places, such as the throat, stomach, ear, and the female reproductive organs.

Lasers have also been effective in removing some birthmarks and tattoos, helping to promote the healing of ulcers and wounds, and removing tooth decay. In some types of brain surgery, lasers are indispensable.

Because laser light can be focused by lenses to a very fine point, it can vaporize small segments of a single cell. Thus surgery on tissues as microscopic as chromosomes can be performed with lasers.

When lasers were developed two decades ago, some scientists thought they would revolutionize medicine. But today, even though lasers have taken on many vital surgical uses, the revolution has not

A surgeon trains a "cold" laser against his hand. This harmless, visible beam, produced by a helium neon laser, is used as an aiming light for a carbon dioxide surgical laser. Carbon dioxide laser beams are invisible. Why? (Lahey Clinic Medical Center)

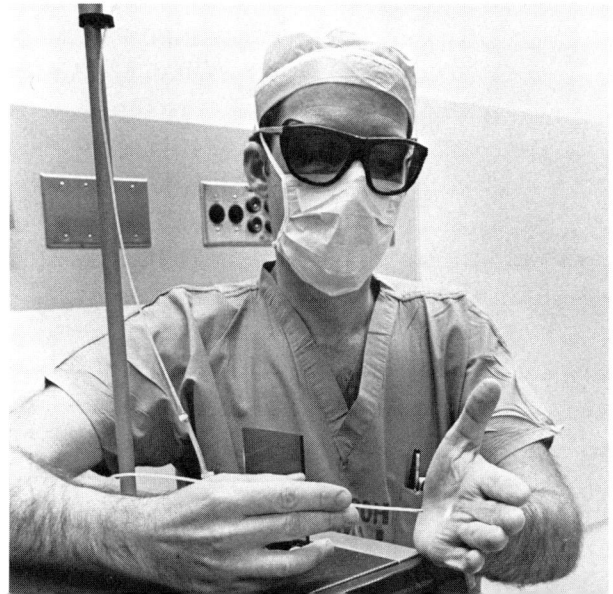

come about. Except for the few specialized operations that have been mentioned, lasers still have yet to gain wide acceptance among doctors.

One reason is cost. A common 50-watt carbon dioxide laser can cost from ten to fifty thousand dollars. Another reason is training. Surgeons have to be specially trained to use lasers. For these reasons, most lasers are found not in doctor's offices but, instead, in large teaching hospitals. But with lasers becoming less expensive and more refined, their future in medicine is bright indeed.

The development of laser as a "scalpel of light" marked a major advance in some kinds of surgery. Here, surgeons remove a tumor from the patient's vocal cords. Using a laser instead of a conventional scalpel enables surgeons to remove the tumor without damaging surrounding tissues—thus preserving the patient's voice. (Lahey Clinic Medical Center)

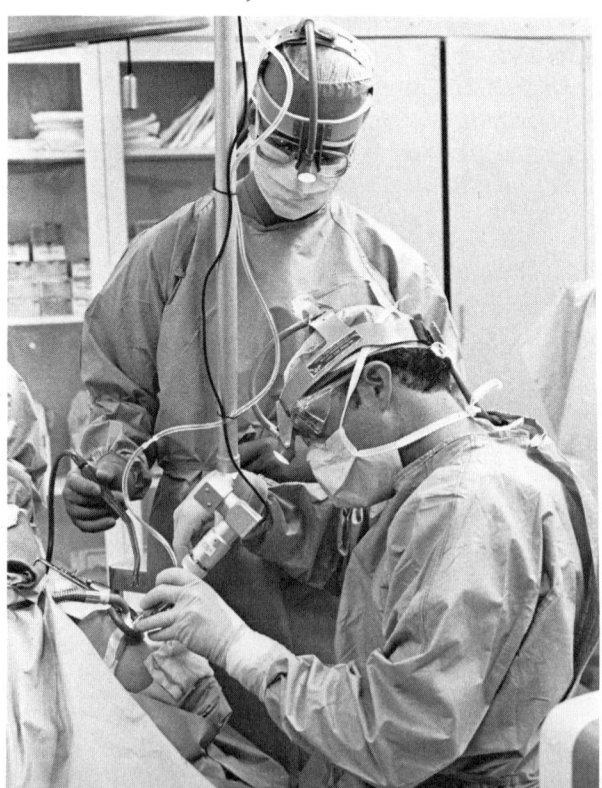

MILITARY USES OF LASERS

In 1898, the British writer H. G. Wells published a novel called *War of the Worlds.* In it, he described jellyfish-like Martians who invaded the earth. The weapons they used were heat rays. In a "flash of light," the rays destroyed anything they were aimed at. Wells's weapons were fantasy, of course, but since then many science-fiction writers have described ray guns in their stories. Heroes and villains alike, in such adventures as *Buck Rogers, Star Wars,* and *Doctor Who,* have used laserlike weapons to punch holes in thick metal, blow up space ships and planets, and paralyze or vaporize their enemies.

Are such striking capabilities even remotely possible with lasers today or in the near future?

Lasers have many exciting uses. But they also have limitations. For one thing, lasers as we know them today might vaporize tiny cells, but they cannot do the same thing to anything much larger. That's because lasers take a lot of energy to operate. More than half the energy used to lase is given off as heat. Our most powerful lasers today can seriously burn a victim, but certainly cannot make him explode. In order for a laser to fire a beam that would kill a person, its power plant would have to be as big as a tank.

In the foreseeable future, ordinary bullets will not be replaced by lasers. The destructive force of bullets comes from their weight and their speed. Lasers have speed, of course. Photons travel at the speed of light. But anything that travels at the speed of light cannot have weight. Photons have no weight. They damage things by burning them. Bullets cause damage or death by tearing things apart.

Despite the limitations of today's lasers, a great deal of research is being conducted on "Star Wars" weapons systems. Before examining some of those systems and the ideas behind them, let's see how lasers are being used today in the military.

Despite the fact that the "fiction" of lasers does not, at least for now, square with reality, lasers do have many military uses today. Low-energy lasers are providing extremely accurate range finding and guidance.

When a gun is fired, it is important to know how

far away the target is. Before lasers were developed, distance was measured by guessing or by taking ranging shots. Both methods were relatively slow and inaccurate. But today lasers can find the exact distance to a target in an instant. This is done by bouncing a laser beam off the target. A calculator then determines how long it takes the beam to go out, reflect off the target, and come back. Since a light beam travels at 186,000 miles per second, it takes only an instant to determine how far away a target is. Laser range finders are being used with antitank weapons and with aircraft flying close-air support.

Laser-guided "smart bombs" were developed for use in Vietnam and are available for use today. Smart bombs are made by fitting regular bombs with laser-seeking heads. When they are dropped from airplanes, the bombs are guided to their targets by laser beams. The beams are fired either by a soldier on the ground or from another aircraft.

Now let's look at the possibilities of using lasers as "Star Wars"–type weapons in space. On the one hand, some scientists say that lasers can be used to destroy missiles and space satellites. On the other hand, other scientists say there are big problems that must be overcome before such weapons can be made reliable.

Certainly the idea of laser space weapons is interesting. A laser shot from 1,000 miles away at a missile that is traveling six times the speed of sound (4,400 miles per hour) would reach its target after the missile had traveled only nine feet. If enough laser-

Shootout in space. One day a space laser satellite may look like this, hurling a killer-beam across hundreds or even thousands of miles of space. But for this scene to become a reality, many problems have to be overcome. (Defense Department)

firing weapons were put into orbit, they could, in theory, destroy all missiles launched from land or sea.

But let's look at the problems. Remember, lasers can do damage only by focusing on a tiny spot. A laser beam would destroy a missile not by blasting it out of the sky, but by burning a small hole in the nose cone. The beam would fry the missile's electronic circuits, knocking out the guidance and warhead systems. The missile would then fall harmlessly back to earth. But in order to burn through the nose cone, the laser beam would have to lock onto the skin of the cone for a moment.

Remember, at this critical moment, the missile is moving swiftly. The beam, which has traveled hundreds and maybe thousands of miles, must remain trained on a single spot. If the beam moves off that spot, its energy will be spread, and it may not heat up the metal enough to burn through. Achieving this kind of pinpoint accuracy is a major problem that will take many years to solve.

But it is not the only problem. Today's satellites are run off solar cells and batteries that are about as powerful as a kitchen toaster. Laser weapons will need a lot more power, at least 100 megawatts, enough for a city. A power plant that would generate that amount of electricity would be enormous.

A laser weapon would need enormous power. Putting such a power plant in space would be very difficult. One answer is to build the laser not in space, but on earth. (Defense Department)

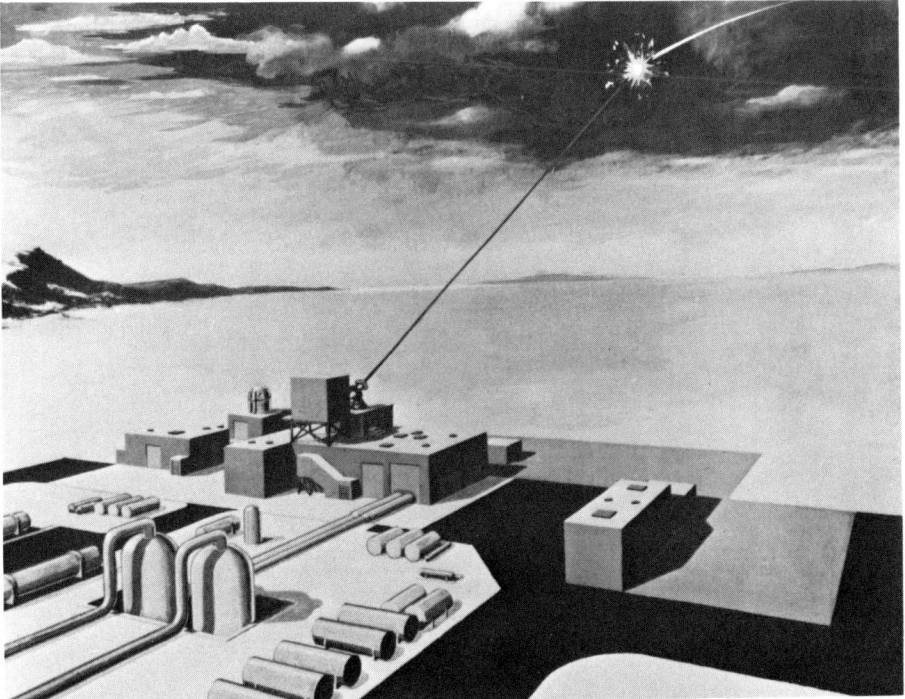

Given today's standards, it would take several hundred acres of solar panels to power a laser weapon. Chemical fuel to generate a weapon would last only a few minutes. Scientists are studying the possibilities of using nuclear power plants. But there are problems with that solution, mainly the tremendous heat the plants would generate, and really big power plants won't be ready until the twenty-first century.

Finally, even if the weapons and power plants are put into orbit, there are more problems. They could be easily tracked and shot down. And missiles themselves might be made safe with special coatings that would turn away laser beams.

But despite these problems, many scientists say that antimissile laser weapons still can be made. They say they might be made with lasers of short wavelengths, X-rays, and gamma rays. Right now a great deal of research is being carried out to try to make these seemingly unreal weapons a reality.

·8·
COMMUNICATIONS

Alexander Graham Bell is famous for inventing the telephone. The telephone uses electrical waves to carry sound over wires. But Bell said the telephone wasn't his most important invention; that particular invention is hardly known at all. He called it the photophone. It used light beams, not electricity, to transmit voices. The trouble was that the photophone, which relied on mirrors and sunlight, did not work on cloudy or foggy days. Bell eventually packed it away and it was forgotten.

But one hundred years later, in 1980, it was unpacked and made to work. Bell's photophone showed that his ideas concerning communication and light were far ahead of his time.

Alexander Graham Bell in 1912. (AT&T Bell Laboratories)

Bell's photophone. It worked by bouncing sunlight from a reflector through a lens. Speech made the sunlight vibrate. The light was then beamed to a receiver. (AT&T Bell Laboratories)

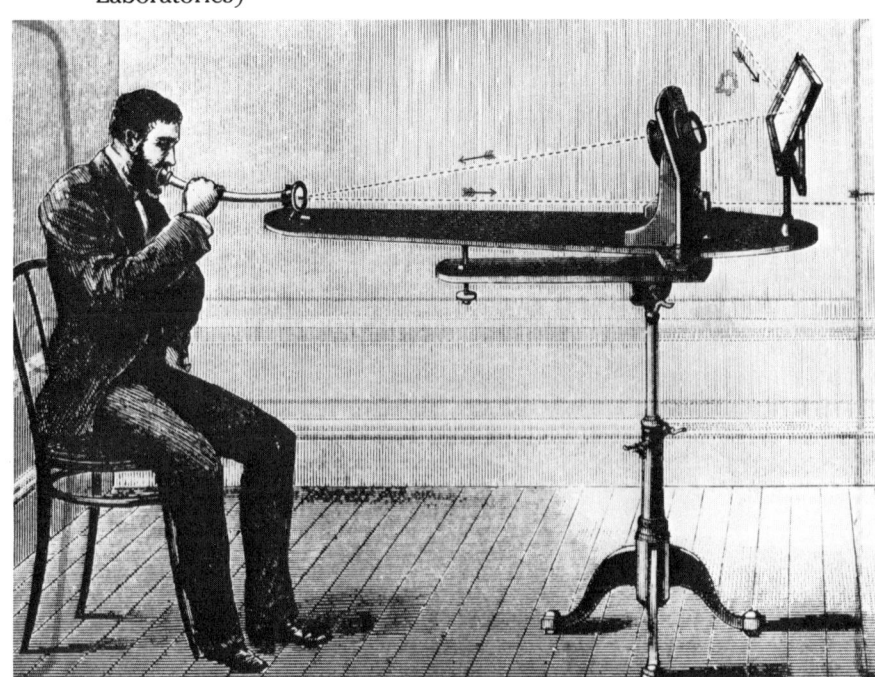

At the receiver, the light struck a selenium detector. The detector vibrated in response to speech. These vibrations were transformed into electrical current. The current recreated speech through a telephone receiver. (AT&T Bell Laboratories)

Light can carry a tremendous amount of information. Remember, light is a form of electromagnetic energy. It travels in waves. Because light waves have a high frequency, they carry much more information than telephone, radio, or television signals.

For instance, a single telephone wire can carry 60,000 bits of information per second. An FM radio can carry 250,000 bits per second. A television carries over 6 million per second. But a laser beam can carry some 100 billion bits per second! In theory, one laser beam can carry all the information that travels across all the world's radio channels!

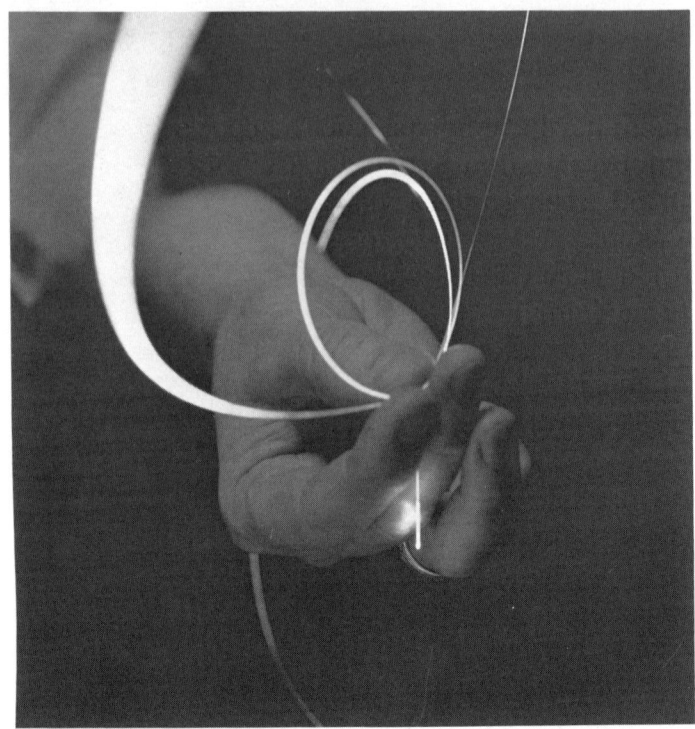

An optical fiber, filled with laser light. (AT&T Bell Laboratories)

And the problems of fog, clouds, and rain that stumped Bell have been overcome today. Laser beams can be sent over glass wires or fibers as thin as a hair. One fiberoptic line can carry about a thousand phone calls in an instant.

◆ 9 ◆
LASERS IN INDUSTRY

When lasers were first developed in the early 1960s, researchers used a simple test to measure their power. They put razor blades in front of the lasers. Focusing the beams to an intense pinpoint with a lens, they burned holes in the blades. Laser power was then measured informally as "Gillettes"—or the number of blades a beam would cut through.

These demonstrations helped create the popular idea that lasers could make holes in or vaporize just about anything. Is this idea correct?

Over the years, lasers, despite the limitations we have discussed, have been doing a variety of jobs in industry. In fact, they are taking on many more

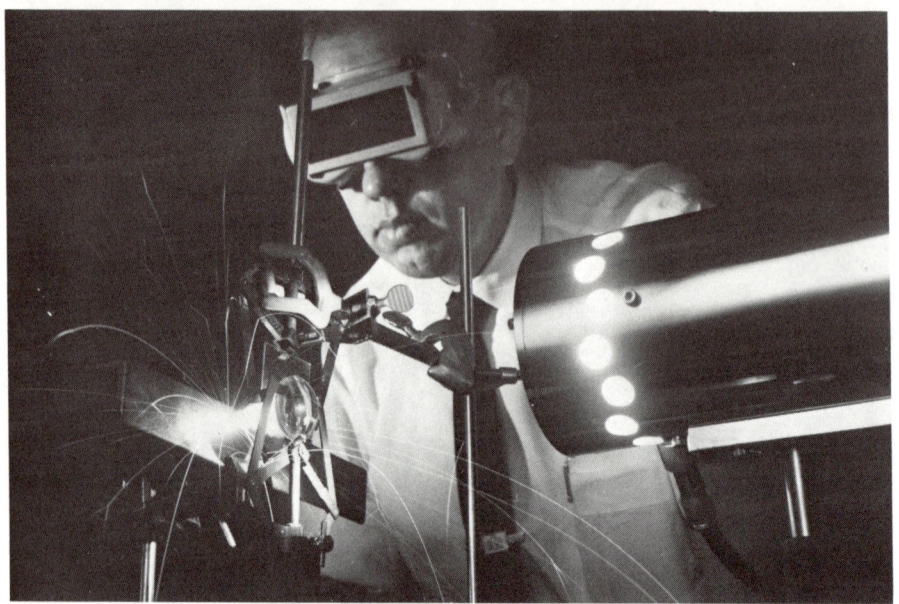

Cold-rolled steel being vaporized by a powerful ruby laser. (General Electric Research and Development Center)

Light from a stationary laser is fed through a glass fiber to a robot twenty-five yards away. The robot cuts, welds, and drills—all with laser beams. (General Electric Research and Development Center)

kinds of tasks than were ever thought of in those early days.

Take drilling, for example. Lasers can actually vaporize the material being drilled. That means that no drill touches the material. You don't have to use drill bits, which can break and need cleaning and replacing. Pulsed lasers can build up enough power to drill the hardest substances, diamonds and titanium, for instance.

Lasers can also do a good job of drilling very soft substances. Since the mid-1960s, carbon dioxide lasers have been making the very tiny holes in rubber baby-bottle nipples. Recently, lasers have been used to drill small holes in aerosol cans and also in rocket nozzles to ensure that fuel is evenly burned. Lasers are the most effective means of drilling ceramic circuit boards for computers and computer games.

But drilling with lasers has disadvantages too. The cost can be very high. Simple laser drilling units run about ten thousand dollars. More complicated and powerful units can cost more than half a million dollars. In most cases, it costs far less to drill holes with machines and bits. In addition, a hole drilled by a laser has a ring of metal surrounding it. Laser-drilled holes are shaped inward, or countersunk, rather than straight. Finally, lasers cannot drill materials that are highly reflective. Why?

In addition, there are safety considerations. When metal is vaporized, it mixes with the air we breathe. Nobody knows what the long-term consequences will be of having vaporized metal in the air.

This hole, about one-tenth the diameter of a human hair, was drilled by a laser. It has the characteristic metal splashing around it. (AT&T Bell Laboratories)

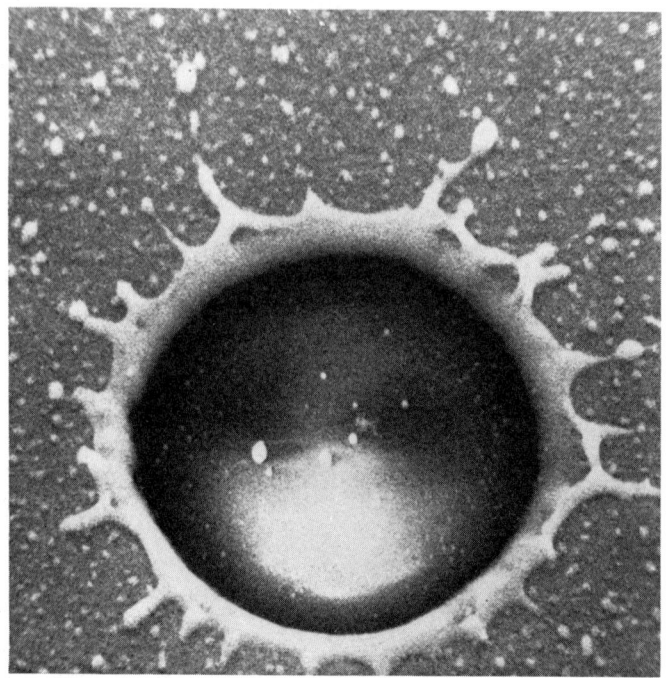

Lasers can be used to cut materials too. They do this in the same way that they drill—by either melting or vaporizing. But they can cut by another kind of process. Pure oxygen is made to flow across the material. The oxygen is then heated with a laser and catches fire. Thus the laser doesn't cut, the burning oxygen does. These types of lasers can cut almost any material, including titanium, which would ruin the blades of most regular cutting tools.

Lasers have created new ways of welding materials together. Welding is a way of uniting two pieces of metal by first melting their edges. With glass tubes that can direct laser beams around corners, welds can now be made in places that would be impossible to reach with conventional devices. Furthermore, with laser welding, impurities are drawn to the surface, making the weld stronger than if it were done the regular way. Because lasers can generate intense heat, thick pieces of cold-rolled steel (up to an inch thick) can be welded quickly.

Lasers are increasingly being used to manufacture computer circuit boards. These boards are be-

Fig. 12 Laser beam drilling metal.

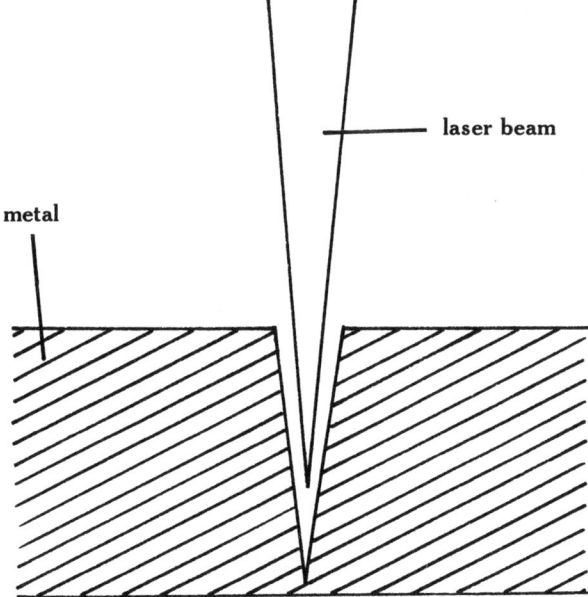

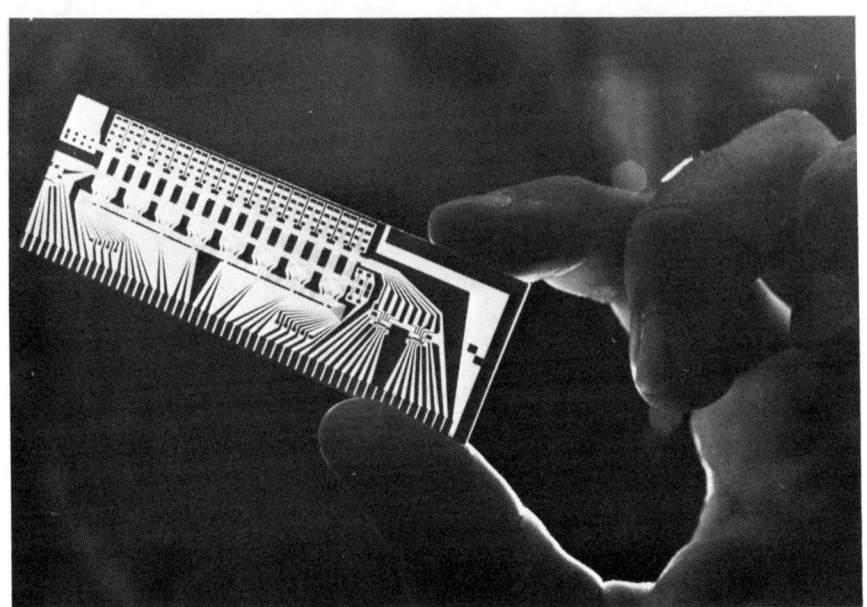

An electronic circuit pattern machined by a laser. (AT&T Bell Laboratories)

coming smaller and more complicated. First, computer-controlled gas lasers draw the patterns of the circuitry. Then a computerized laser system can solder as many as forty joints per second with as many as one hundred connections per square inch—ten times faster than with conventional soldering techniques. This means that in the years to come, computer circuit boards will, with the help of laser solder, be made better, cheaper, and faster than ever before.

Here are some more uses of lasers in industry.

- They can strip insulation from coaxial cables. Coaxial cables carry telephone and television signals. Each cable is made of many wires, so it can carry many signals. Each wire is protected by plastic insulation. Stripping the plastic off by mechanical means is difficult, but a laser can do it quickly and easily. The plastic absorbs the energy of the beam, and the metal reflects it. Thus the laser can strip away all the plastic, leaving only the wire. A hand-held laser wire-stripper is a standard piece of equipment on the space shuttles.

- Pulsed lasers, guided by computers, are writing identifying marks such as serial numbers, trademarks, etc., on a variety of goods.
- Computerized carbon dioxide lasers are cutting out patterns for suits from bolts of cloth. Fifty suits per hour can be cut out by twin CO_2 lasers.
- Lasers are removing ink from paper without damaging the paper. How can lasers do that?
- Pulsed ruby lasers are drilling holes in watch jewels and in industrial diamonds.

⋅10⋅
MEASUREMENT BY LASERS

On the door of the laboratory is the sign LASER EXPERIMENT IN PROGRESS—DO NOT ENTER. The Bell Laboratories physicist is sitting behind a protective wall. He warns visitors, "Whatever you do, don't look at the laser." Light from the laser, which is pulsing in the center of the room, flickers in cherry-red splashes on the wall.

This is a very special laser, located at Bell's New Jersey labs. There is no other like it in the world. It is producing the world's quickest flash of laser light.

It begins and ends within 30 femtoseconds.

Since the laser was first developed, shorter and shorter pulses of laser light have been triggered by

The Bell Laboratories' laser that creates pulses of light measured in the billionths of a second. Such incredibly short pulses enable us to better understand subtle physical and chemical actions. (AT&T Bell Laboratories)

researchers. Lasers were made which reduce pulse duration first to the nano (one-billionth) of a second, then to a pico (one-trillionth) of a second. Today with the laser at the Bell labs, light pulses are measured in femtoseconds, or quadrillionths of a second.

A Bell researcher explains just how fast that duration of time is. "In a second a pulse of light can travel almost to the moon. But in 30 femtoseconds, light travels only about one-tenth the thickness of a human hair."

Just why has such an ultrafast, multi-million-dollar laser been developed? To answer that question, let's remember what we examined in chapter 3. Light is a packet of energy, in the form of a photon, that is emitted from a stimulated electron. The time between stimulation and emission is incredibly short. Many other key events in matter take place in a similarly short period of time, such as atoms colliding, chemicals reacting, and electrons vibrating. Interestingly, these events are very difficult to measure. This is not only because they happen quickly and invisibly, but because events in that small realm cannot be accurately predicted, as can events in our visible world.

Yet these tiny events are tremendously important. Their sum makes up the world as we know it. Lasers, such as the one at Bell Laboratories, allow us to explore and measure these events in this way: The short, coherent light pulses excite atomic actions. Since laser pulses are extremely accurate, the changes in light that the atomic actions create can be measured and examined. In this way, lasers can give us glimpses into worlds conventional instruments cannot explore.

Lasers can detect and measure things as large as pollutants in smokestack emissions and as small as . . .

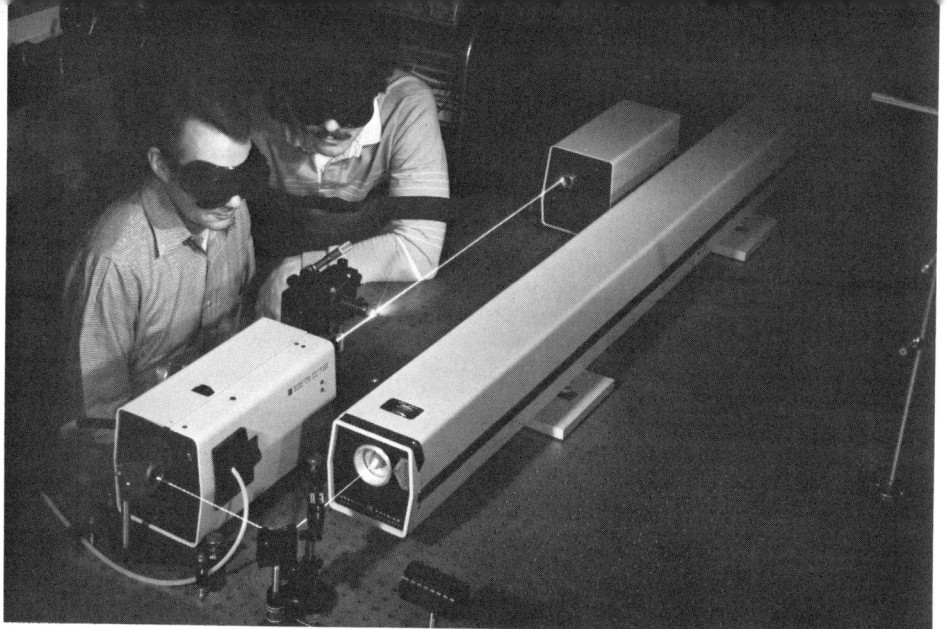

. . . the emission lines of atoms. Here at the Oak Ridge National Laboratory scientists are analyzing minute chemical reactions with a dye laser. (United States Department of Energy)

Things we learn about what happens in matter during very short spaces of time have led and will lead to breakthroughs in microelectronics and communications.

The Bell Lab laser is a vivid example of one important feature of lasers. They can measure with amazing accuracy and speed.

Other measuring lasers are doing other kinds of tasks. For instance, lasers are being used to provide surveying lines and straight lines for buildings. Gas lasers diverge less than one part in a thousand and are widely used for alignment in large construction, laying pipe, drilling tunnels, and guiding heavy machinery.

Lasers are also being used in light radar. Conventional radar depends on sound waves. But light radar travels faster and gives clearer pictures than sound radar.

The first astronauts to land on the moon placed a reflector there. A laser beam fired from earth and bouncing off the reflector has measured the moon-to-earth distance to within 15 centimeters. Furthermore, the drifting of the earth's continents has been measured by beaming lasers against the reflector from different parts of the earth.

Earthquakes can be more accurately reported with a laser geodynamic satellite. Mirrors in the satellite reflect laser light so that even the smallest movements in the earth's surface can be instantly noted. This may eventually lead to accurate earthquake prediction. Also, surface motion can be read by lasers across a fault. Three lasers are operating in Southern California 25 kilometers (15.5 miles) from the San Andreas fault. They can detect strain in the earth's surface as one part in a billion.

Surveying too has been helped by lasers. For instance, surveying the Grand Canyon the regular way took one year and upwards of a hundred men. But lasers did it in three days.

Lasers trained around a ring can serve as a gyroscope with no moving parts.

Lasers can scan for defects in materials at supersonic speeds. Aided by a computer, lasers can make swift, accurate, and complicated cuts in wood. Finally, lasers are being developed that can synchronize clocks to within billionths of a second.

♦ 11 ♦
HOLOGRAMS AND LASERS

A hologram is a three-dimensional photograph. Look at one, and you might think your eyes are playing tricks on you. You can see into it, just as if it were a real-life image. What's more, you can actually look at the sides of the objects in the photograph by moving your head. Yet if you touch it, you'll touch a flat surface.

The word hologram comes from two Greek words: *holos,* meaning whole, and *graphos,* meaning message. Thus *hologram* means "the whole message."

The basic idea of holography was developed in 1947, but it took the invention of the laser to make that idea a startling reality.

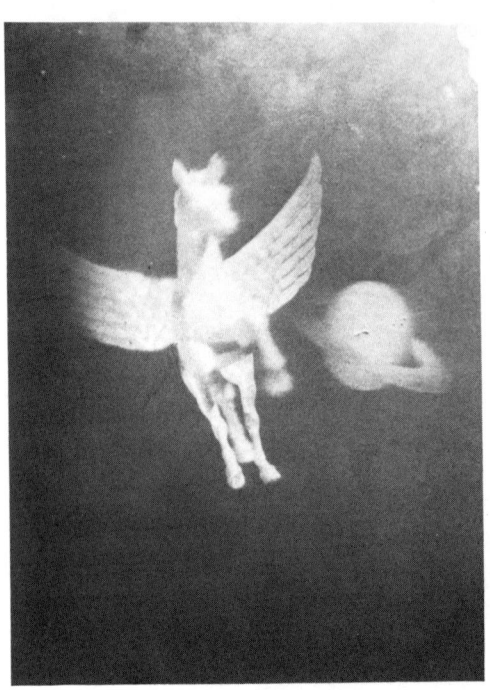

Two photographs of one hologram. Saturn is seen hiding behind this hologram of Pegasus when the angle of view is changed. (Jason Sapan/Holographic Studios, NY)

How can such a photograph be made? In order to answer this, let's first examine once more how we see things. For this purpose, we will be discussing the wave aspects of light.

When we see an object, we are actually seeing a wave-front of light coming from the object. The wave-front pours into our eyes, and our brain interprets it. The idea behind a hologram is to recreate that same wave-front on film.

As we know, a photograph taken by a camera does not reproduce the wave-front exactly as we see it. Instead, it produces an image that has no depth. Here is the reason. Light waves are made up of two aspects. One aspect is *amplitude*. The other

Fig. 13 Amplitude and phase of light waves.

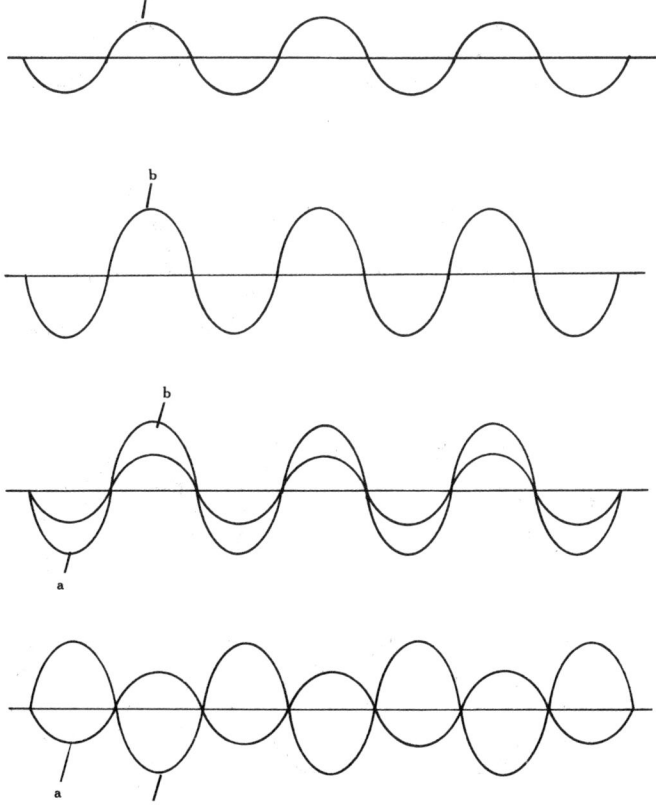

aspect is *phase*. Amplitude is the distance between the top and the bottom of each wave. The greater the distance between the top and the bottom, the greater the amplitude. Amplitude deals with the intensity or brightness of light. When the amplitude of light waves is large (that is, when the distance between the tops and bottoms of the waves is great), the light is bright. When amplitude shrinks, light dims, becoming less intense.

A camera photograph records amplitude. In other words, the film shows the intensity of the light. It becomes blacker when intensity is greater. This is the basis of normal photography. But normal photography does not capture phase. It captures only amplitude.

Fig. 14 Making a holograph.

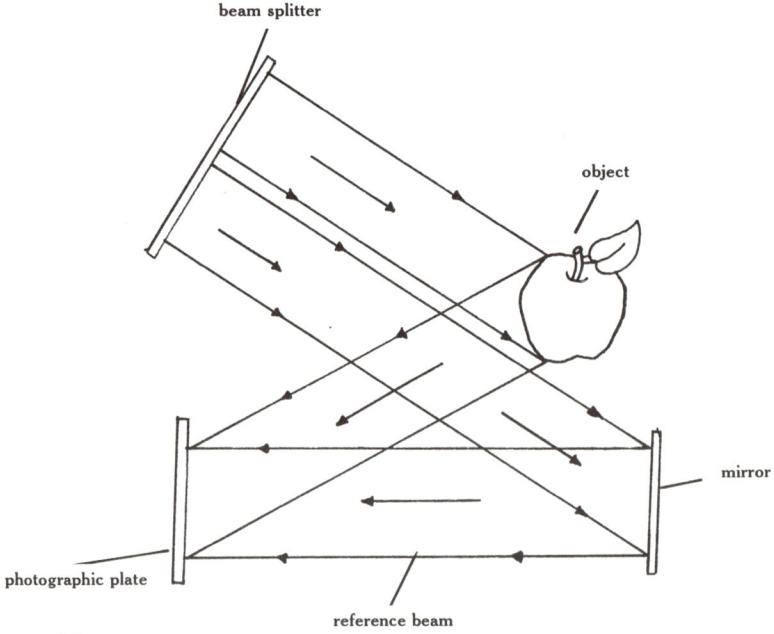

What is phase? It's simple. Phase is the *shape* of the light wave. In order to recreate a wave-front, we must capture both amplitude and phase—or both the object's brightness and its shape. This is where lasers come in.

A hologram is a record of both amplitude and phase. It is a record that is made without a lens. What is needed is a beam of laser light. The beam is split into two beams. One beam bounces off a mirror and then shines on the object. The other beam shines directly on the plate. The combination of both beams creates the holographic image.

But why can't normal light be used to create a hologram? It can, under special circumstances. Lasers, however, can do the job much better. Here's why. Remember what we said about the characteristics of laser light. It differs from normal light in that its waves are coherent. In other words, they are in step with each other. If two beams of incoherent light fall on the plate, the light merely gets brighter.

Fig. 15

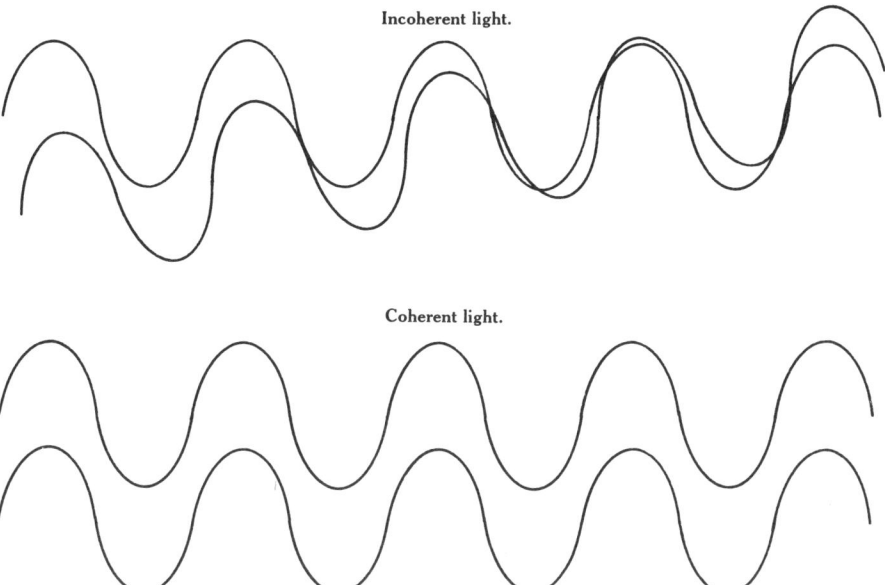

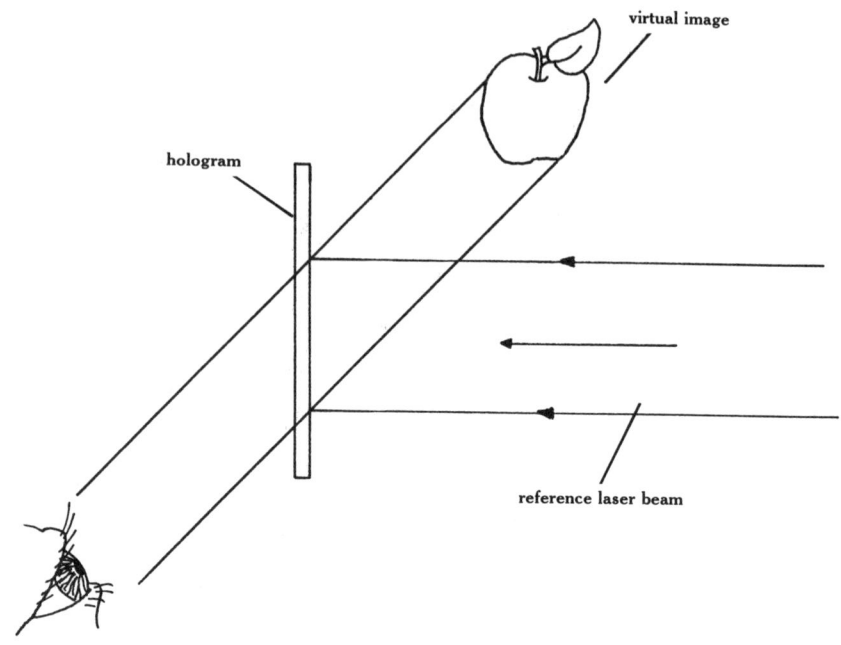

Fig. 16 Viewing a hologram.

But when the two beams of laser or coherent light fall on the plate, an interesting thing happens. By a special process called *interference,* the very shapes of the light waves are frozen on the plate. When we look at the plate, we see the image both in brightness and in depth. We see a hologram.

Here are some interesting things about holograms.

- Unlike a conventional photograph, a lens is not needed to make a hologram. This is because the photographic plate receives the exact wave-front that we see.

- A hologram can be scrambled into a code. Only the person with a decoder can recreate the true image.

- A cylindrical hologram can be made that will reveal all the outer surfaces of an image. A person can walk around the cylinder and see its front, sides, and back. Turn the cylinder upside down and new views of the image will appear.

- Any section of the hologram can be used to recreate the whole image. That's because the hologram repeats the same wave-front many times across its entire surface.

- The image that appears on the plate is, unlike the conventional photographic image, always positive.

- Double-exposed holograms are being used to test the strength of things, from old paintings to aircraft tires, without destroying them.

- Holographic microscopes offer biologists unique opportunities to make three-dimensional studies of bacterial movement and cell and tissue growth.

There are, of course, limitations to holograms. Making them requires special equipment, including special dampers to keep vibrations to a minimum. The equipment certainly isn't portable like a camera. Though holograms can be made colored like a rainbow, most of them are a single color, usually that of the type of laser used. Their tone is grainy; and in general, large objects cannot be recreated in a hologram because of limitations of the equipment.

Holography is in its earliest stages. Today researchers are attempting to use lasers in developing full-color holograms as well as moving-picture holograms and television holograms.

Holograms can provide accurate stress testing for many different kinds of materials. A laser-hologram pattern has revealed a crack (middle left) in plastic. (AT&T Bell Laboratories)

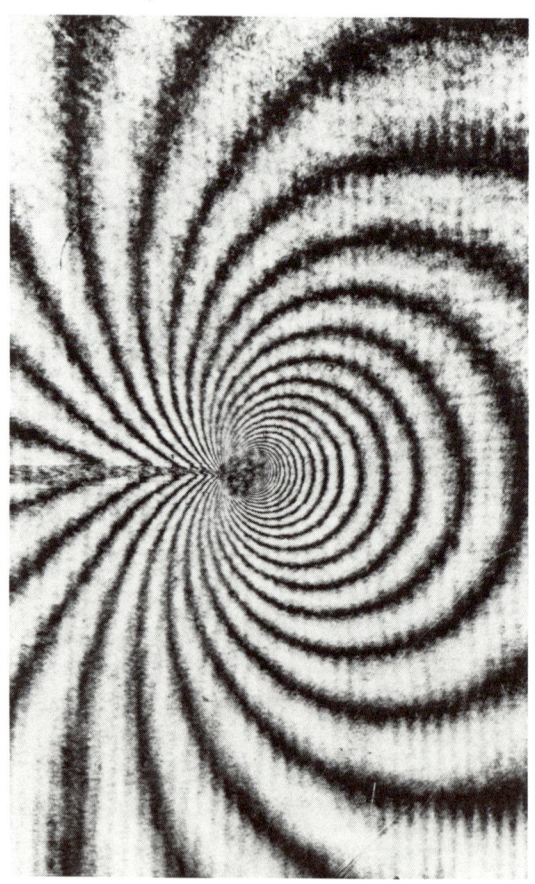

12

A GARDEN OF LASER DELIGHTS

Shortly after the first laser was fired, a James Bond movie came out that featured a laser as an instrument of death. You may have seen the movie, *Goldfinger*. James Bond was strapped to a steel table and the laser beam ripped a flaming, smoking path toward him. That movie created a popular view of lasers. We might call it a hard view. Lasers were viewed primarily as burning, blasting tools of destruction. But over the years, laser technology did not take the hard route. The most important uses of lasers have turned out to be not hard, warlike ones of burning and blasting, but instead the "soft," peaceful ones.

We have already explored some of these soft uses of lasers in medicine, communications, measurements, photography, and industry. Most of these uses do not have a direct impact on our lives. But there are many soft uses of lasers that touch our lives daily. We'll talk about some of those uses in this chapter.

Pick up a food package, and you'll probably find a striped label printed on it. That label is the result of laser technology. It is called a Universal Product Code label and is designed to be read by a laser. The laser, usually a low-power helium neon laser, scans the stripes. The coded stripes make changes in the laser beam. Those changes are recorded by an optical detector. The detector sends signals of those recordings to a computer. The computer finds the price of the item. It records the item and its price on the customer's tape. The computer also makes changes in the store's records. It subtracts that item from the total number for sale in the store. It's even possible for the computer to subtract the item from the total number for sale in the region or nation.

A laser can read in other ways too. It can scan letters, symbols, and dots. Because a laser beam can be focused to a very small pinpoint—.001 mm (or .004 inches)—it can scan small areas quickly in order to read at high rates of speed. But lasers do not only read. They can write too. Many major newspapers in the United States and Europe are being printed with lasers.

Laser printing is done this way. A low-powered beam (usually helium neon) scans print and photos that have been pasted on a board. The information picked up by the laser is then transferred by a computer to another laser. This laser is powerful. It reproduces an exact copy of the information on a special plate. The information is then transferred from the plate to the pages of the book or newspaper.

Laser printing has a number of advantages over other types of printing. One is speed. Upwards of twenty thousand lines per minute can be printed by a laser. Another advantage is that the printing can take place simultaneously at distant locations. For instance, information picked up by a low-powered beam in New York can be transmitted immediately via satellite to London. There, the more powerful laser reproduces the information on the plates. The newspapers carrying the same pages then come out in the streets of New York and London at the same time. Another advantage to laser printing is that it eliminates some expensive and time-consuming steps that are involved in conventional printing.

Lasers are being used by NASA to provide exceptionally clear photographs that are taken in space. The photographs are converted to electronic signals which are transmitted to a laser system on earth. The laser can write up to twenty thousand separate pieces of information on a 2.41-centimeter-long (10-inch) line. This unequaled ability to write so finely creates very sharp photos.

Using techniques similar to those that allow them to read and write, lasers can also store vast amounts of information in tiny spaces. For instance, a laser can store some ten billion bits of information

On an optical disk such as this, the entire Encyclopedia Britannica or 50,000 still photographs can be recorded and stored for instant retrieval. Optical disks, as distinct from video discs, are smooth rather than grooved. A strong laser records information by "pitting" the disk. A weaker laser then can read the pits. (RCA)

on a single side of an optical disk. The information is put on the disk digitally by a high-powered laser beam, and the beam burns holes in the light-sensitive coating of the disk. The digital information is then "read" by another, less powerful laser while the disk spins rapidly. The great storage capacity of laser memory systems is helping provide animated sequences in the new generation of video games.

Lasers are also being used in some video-disk systems. In this case, the laser beam acts much like the old phonograph needle does on a record. It picks up information that is stored in the tiny grooves. But since the information is being transmitted by light instead of metal, there is no wear on a laser video disk.

• 13 •
LASERS IN ART AND ENTERTAINMENT

Artists of all eras have tried to represent changing patterns of light in their work. With the advent of lasers, a whole new realm of art has opened up. Laser light itself can be used as an artistic tool. It can etch and sculpt a variety of materials. When it is used with revolving mirrors, it can create dazzling displays. Many laser shows are put on with music. Music controls the movement of the laser beams and mirrors to present an ever changing, multicolored show. Many rock groups, led by the British band The Who, have added laser shows to their performances.

Three-dimensional ribbons of laser light swirl across a revolving screen to create ever-changing laser sculptures. The lasers respond to electrical signals obtained from any stereo music source such as FM radio, organ tones (used here), or taped music. (AT&T Bell Laboratories)

◆ 14 ◆
LASER SAFETY

So far, lasers have proved to be fairly safe. Since their invention more than two decades ago, only about twenty eye injuries resulting from their use have been reported to the federal government. Several physicians have trained low-power lasers against their own skin every day over a period of years with, at least up till now, no ill effect. However, this does not mean that lasers are not dangerous. They can be very dangerous indeed. They can cause blindness, skin damage, and, if the vaporized by-products are inhaled, damage to the lungs and other organs.

Their primary danger is to the eye. Remember that light enters the eye through the cornea and then forms images on the retina. But the light we speak of is normal or incoherent light. When laser or coherent light enters the eye, its narrow beam can burn the retina, causing blindness. This can happen not only when you look directly at laser light, but also when it is reflected off a mirror or shiny surface. Laser beams that can't be seen by the naked eye, such as carbon dioxide lasers, can be especially dangerous, since it may be difficult to know where they are aimed. Special laser-safety glasses should be worn when working with a laser.

Back in chapter 6, we explored how lasers are being used as scalpels in some kinds of surgery. People who work with high-powered lasers wear gloves and can use creams for protection. So far, there has been no evidence linking laser light with skin cancer. But the long-term effects of exposure to coherent light are still being studied.

When powerful lasers are used on metals and plastics, ventilation systems and masks must be used to protect against inhaling the vaporized materials.

Although lasers have been around since 1961, they are still in the very early stage in regard to our knowledge of their effects on tissues. Extensive research is being carried on to examine these effects and help set laser safety standards.

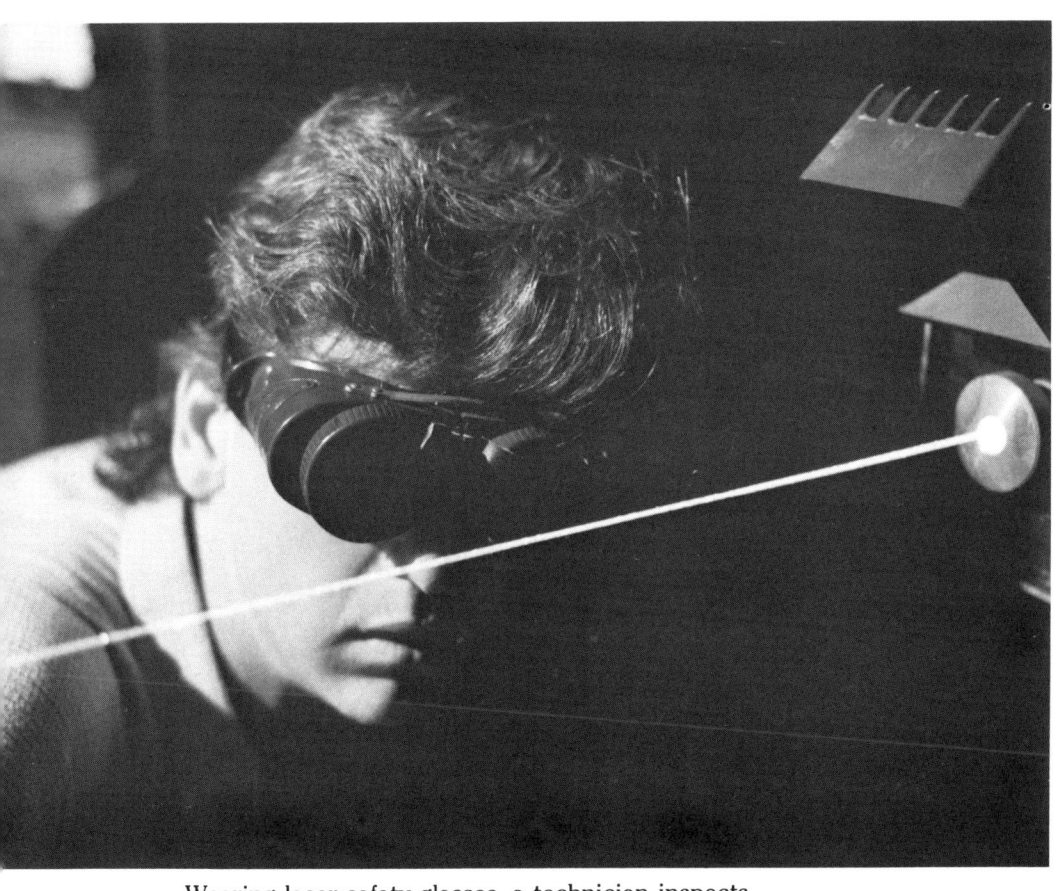

Wearing laser-safety glasses, a technician inspects a beam being fired from an argon laser. (Jason Sapan/Holographic Studios, NY)

THE FUTURE OF LASERS

Interstellar communication. Genetic engineering. Cold laser treatment of paralysis. Laser propulsion. Unlimited energy from water. These are some of the possible future uses of lasers.

One of the most vigorously pursued ideas is the attempt to get energy from water. It is called *atomic fusion*. If its promise is fulfilled, humankind's energy problems will be solved. For it would mean that energy could be derived from the earth's most abundant resource. And unlike atomic energy that comes from *fission*, fusion would create—in theory at least—few if any hazardous wastes.

In order to understand what fusion is, let's first examine fission. Fission comes from a Latin word that means splitting or dividing. When atoms are split, a large explosion results. This happens because atoms are held together by a force that is extremely powerful. The bomb that destroyed Hiroshima was the reaction caused by the splitting of only about ten atoms.

But fusion is the opposite reaction. It is the combining or "fusing" of the elements of two smaller atoms to form a larger atom. When this combination is made, energy is also released. In fact, a great deal more energy is released through fusion than through fission. The sun and stars are powered by fusion. A hydrogen bomb is a fusion reaction.

All the world's atomic power plants use fission power. But there are problems with using this kind of energy. It produces highly radioactive wastes. It uses fuel that is found in limited supply.

But fusion would not produce a great deal of radioactive waste. There are scientists who say that fusion plants may be built that will produce no such waste at all. Its fuel, which could be seawater, is all but unlimited. Yet fusion power is still a long way off. Experts claim that fusion generators will not be operating until the next century. Most scientists agree that solving the problems of making it practical is probably the greatest engineering challenge of all time.

Lasers are a vital part of the challenge. Here's why.

In order for a fusion reaction to take place, three things must happen. The fuel must be heated to

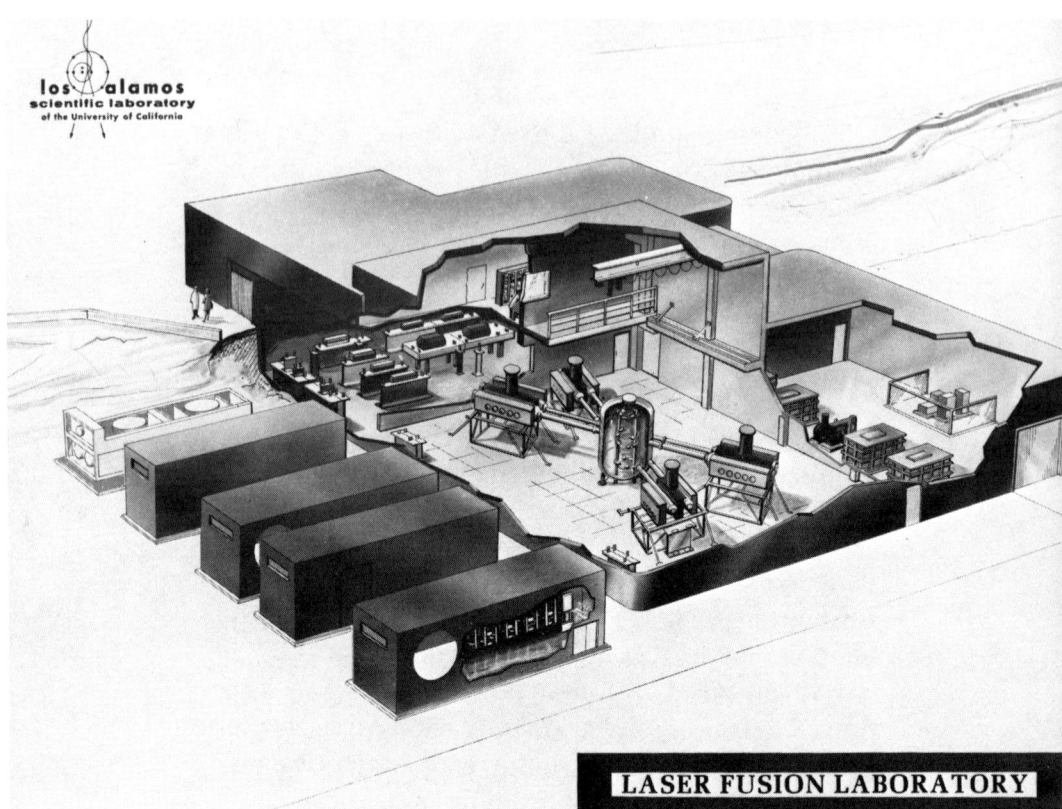

LASER FUSION LABORATORY

An artist's drawing of the basic workings of a laser fusion laboratory. Enormously powerful lasers are positioned around a target chamber. The lasers fire beams against a tiny fuel pellet in the chamber. If the lasers are powerful enough, the pellet should then "implode," triggering a fusion reaction. (Los Alamos National Laboratory)

very high temperatures. It must be compressed, or made much smaller. Then it must be held in that hot, compressed state for a specific length of time while fusing is occurring.

Fusion reactions in stars are created because the enormous gravitational forces compress their matter to a very dense mass.

Fusion reaction in a hydrogen bomb takes place by making fission—or atomic explosion—compress matter.

Clearly, we cannot duplicate the gravitational forces of stars here on earth. Also, A-bomb explosions can't be used in power generators. So there has to be another way to create the right conditions that would make fusion happen.

So far scientists have come up with two possibilities. The first is called a magnetic confinement. Gas is heated in a special doughnut-shaped container. A magnetic field is used to try to keep the gas away from the walls of the container, for the gas is so hot that it can burn out the walls. But so far the gas has proved to be a slippery fish. The magnetic fields have not been strong enough to contain it.

The second possibility is laser fusion. The idea is to zap a pellet of fuel with highly powerful laser beams. The fuel pellet would be about the size of a grain of sand and composed of a special kind of hydrogen called deuterium. The pellet would be encased in a gold jacket. The gold jacket allows even burning when the laser beams strike it. The laser itself would be a tremendous device. It would have to send scores of trillions of watts of power into the pellet in about 0.2 billionth of a second. The fuel would then *implode*—or explode inward—and fusion would result.

To date, some large lasers, several costing many millions of dollars, have been built to help make fusion a reality. Even bigger lasers are being planned. But no laser has yet been built that has the

Fusion fuel pellets are very small. Their sizes range from those seen against a quarter to this pellet framed in the eye of a needle and lying on a human hair. (Los Alamos National Laboratory)

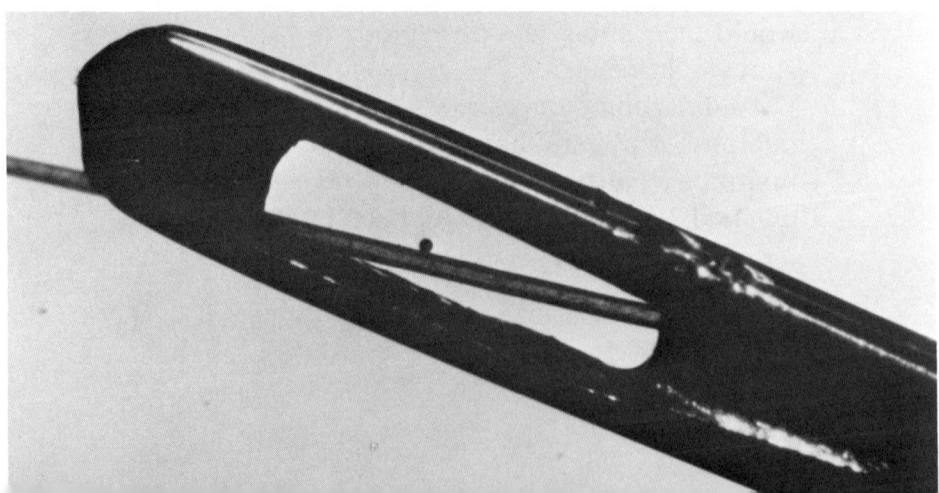

power to create a fusion reaction. Scientists say it won't be until the next decade or even the next century that one will be completed.

While lasers give promise of solving energy problems here on earth, they also hold out the dream that we may be able to communicate with other star systems. Recently, a group of scientists found that a laser is working in the skies of Mars. It is a carbon dioxide laser. It is driven by natural phenomena. Molecules of carbon dioxide in the Martian atmosphere are being stimulated into population inversions by photons of light from the sun. (See chapter 3.)

But the natural output is not as narrow as laser beams created here on earth. To make it narrow, scientists have suggested that two huge mirrors be put in orbit around Mars. The mirrors would then bounce the light back and forth and create an intense laser beam. That light, which can be dimmed or brightened, could then be aimed at distant stars. To an observer in far-off regions of the Milky Way, the beam would be some seven hundred times brighter than the sun.

Lasers have future uses not only in the vastness of space, but also in the microscopic world of cells and genes. Genes are tiny elements located in our cells. These elements determine the makeup of our bodies: the color of our hair and eyes, our sex, our height, etc. Many human maladies are caused by defective genes. But some of these disorders may be corrected or prevented by making changes in the genes. The trouble is, genes are tiny and fragile. But lasers, which, as we have noted, can be focused to a pinpoint and can be guided by a computer, may be able to make these changes swiftly, accurately, and effectively.

Genetic engineering will be done with low-powered lasers. But lasers with even lower power may one day be used to help paralyzed people regain movement. These lasers have such low power, they are called "cold" lasers. Researchers in California have noted that treatment with cold lasers has helped reduce pain and promoted some movement in paralyzed limbs. Nobody knows exactly how a cold laser affects tissues—if it really does. But the study that reseachers are now giving cold lasers may eventually have dramatic payoffs.

Lasers that would power space- and aircraft is another idea that is being seriously considered by some scientists. One of the unusual aspects of this idea is that the laser would not be attached to the craft. Instead, it would be aimed from the ground or from a satellite. The beam would then strike the craft and heat up its fuel, and the heated fuel would drive the craft. The main advantage of laser-propelled rockets and airplanes would be fuel efficiency. Researchers who support the idea claim it would save upwards of six times the amount of fuel now used in a jet ride across the continental United States. It may also be fifty times less expensive to blast payloads into orbit.

But laser propulsion—if it ever can be attained—is many years away. Very powerful lasers would have to be used. Their size would be enormous. Their cost would be enormous. But they may be the propulsion of the future.

These are just a few of the possible future uses of lasers. It is too early to tell if we will see them come to pass. But what is certain is that in the years to come, many more uses of lasers will be found. In 1961, nobody thought that lasers would be playing such vital roles in computers, food packaging, and printing. No doubt many of the dramatic future uses of lasers are only dreams today.

GLOSSARY

absorption—takes place when an atom is excited and its electrons jump to more distant orbits. The result is that electromagnetic radiation is reduced.

Al-Hazen—an Arabian scientist of the eleventh century who made important studies dealing with the eye and vision.

amplification—increasing the strength. For instance, the amplification of light means that it is made more powerful.

amplitude—the height of a light wave's crest above an imaginary horizontal line. See **wavelength, frequency,** and **phase.**

atom—a very small particle of matter composed of protons and neutrons and orbiting electrons.

Alexander Graham Bell—American scientist and inventor (1847–1922) who invented the telephone and photophone.

biologist—a scientist who studies living matter.

Niels Bohr—a Danish physicist who in 1913 declared that electrons orbit around an atom not randomly but in specific distances from the atom's center or nucleus.

chromosomes—threadlike particles located in the cells of living things. The chromosomes carry genes which govern heredity.

circuitry—the pathway devices that allow an electric current to make a complete course.

coaxial cable—a cable made up of a tube through which many wires are passed. Coaxial cables are used to transmit a great number of telephone and telegraph signals at the same time.

coherent light—light that does not spread out but remains in a narrow line, and the waves of which are in step with one another. Laser light is coherent light. See **incoherent light**.

computer circuit board—a hard surface that holds the electrical pathways or circuitry of a computer in place.

cylindrical hologram—a hologram placed around a cylinder so that all sides of the subject can be seen.

deuterium—a heavy form of hydrogen.

digital—in computer technology: a whole number used in representing a problem to be solved.

Albert Einstein—a German physicist (1879-1955) who developed the theory of relativity. Einstein first speculated on the possibilities of laser light.

electromagnetic energy—energy that results from the interaction of electric currents and magnetic fields.

electromagnetic radiation—radiation in the form of electromagnetic waves. Radio waves, light waves, X-rays, and gamma rays are all forms of electromagnetic radiation. They differ only in length.

electron—an invisible form of negatively charged electricity that orbits an atom.

ether—the imaginary substance that nineteenth century scientists thought was supposed to exist throughout the universe and be the medium through which light traveled.

femtosecond—an extremely small fraction of a second that can be measured by short impulse lasers.

fission—the splitting of a nucleus of a heavier atom into the nuclei of lighter atoms. The explosion of an atomic bomb is a fission reaction. See **nucleus**, **atom**.

frequency—in regard to the electromagnetic radiation: the number of times waves of energy pass a certain point in a second.

fusion—a thermonuclear reaction in which the nuclei of light atoms fuse to form heavier atoms. The explosion of a hydrogen bomb is a fusion reaction.

genetic engineering—re-making hereditary characteristics by arranging genetic material.

gravity—the force that draws objects toward the center mass of the earth. The result of gravity is weight.

ground state—the lowest energy state of an atom.

hologram—a three-dimensional photograph taken with the aid of a laser or other coherent light and without the use of a lens.

holographic microscopes—microscopes that show three-dimensional images.

Christian Huygens—a Dutch mathematician, physicist, and astronomer (1629–95).

hydrogen—the lightest and simplest element. It is composed of one proton and one orbiting electron.

incoherent light—normal light as opposed to laser or coherent light. Incoherent light spreads out and is composed of many colors. See **coherent light**.

interference—the process in which light waves clash and either cancel each other out or reinforce each other.

interstellar—between stars. Interstellar communication would be communication between stars.

Philipp Lenard—a French physicist who won the Nobel Prize in 1905 for discoveries concerning electrons and light.

light—a form of electromagnetic radiation that reacts with the eye to make objects visible.

magnetic confinement—keeping super-heated gas away from its container walls by magnetic force so that a fusion reaction can take place. The gas is so hot that it would melt the walls.

Theodore Maiman—an American scientist who built the first laser.

maser—an acronym that means Microwave Amplification by Stimulated Emission of Radiation.

James Maxwell—a Scottish physicist (1831–79) who discovered the existence of electromagnetic fields. See **electromagnetic radiation**.

microelectronics—electronics that operates on a very small scale.

microscope—a device that visually enlarges objects that are too small or nearly too small to be seen by the naked eye. See **holograph microscope**.

microwaves—electromagnetic waves of very high frequency.

mode-locking—a way of increasing the power of a laser burst by switching laser light on and off in extremely short pulses.

nanosecond—one billionth of a second.

Isaac Newton—an English philosopher and mathematician (1642–1727). Newton formulated the law of gravity, and he studied the nature of light.

nucleus—the central part of an atom. It contains the protons and neutrons and makes up most of the atom's weight. The nucleus is positively charged, as opposed to the orbiting electrons, which have negative charges of electricity. Plural: nuclei.

optic fiber—a plastic or glass filament through which light is sent to carry information.

optical disk—a disk that can have information stored on it and that is activated by a beam of laser light.

optically pumped—exciting atoms to create a laser action by means of powerful bursts of light, such as are given off by a flashlamp.

orbit—the path one object makes when it moves round and round another object.

phase—a specific state at a given instant in a cycle. The shape of a light wave is determined by its phase. See **wavelength, wave front,** and **amplitude.**

phenomena—facts and actions that can be observed: for instance, the phenomena of light and the phenomena of electricity.

photon—a particle of energy in the form of light.

photophone—a device, invented by Alexander Graham Bell, that sends voice signals by light.

picosecond—one trillionth of a second.

physicist—a scientist whose special study is physics.

Plato—a Greek philosopher who lived from 427 to 347 B.C.

population inversion—a special condition of energy when more atoms are in an excited state than in a ground state. A population inversion is necessary before lasing can take place.

physics—the science that deals with matter, energy, motion, and force.

Max Planck—a German physicist (1858–1947) who made a revolutionary insight in 1900: that atoms emit energy in explosive bursts.

Q Switch—a device that helps increase the power of a laser. With the switch closed, the laser energy cannot escape its cavity. Thus, it builds up tremendous power. Then, when the switch is suddenly opened, an extra-powerful burst of laser light is released.

radiation—energy being released in the form of waves or particles.

range finders—optical devices for estimating the distance between objects.

ranging shot—estimating the distance to a target by firing a small caliber bullet, usually a tracer.

retina—a coating of tiny rods and cones on the inner lining of the back of the eye. Light pours through the lens of the eye and forms images on the retina.

semiconductor—a material, usually a crystal, that does not conduct electricity as well as metal, yet it is not an insulator. A semiconductor can increase its conductivity by being exposed to light or heat.

shortwave—an electromagnetic wave of less than sixty meters. A normal radio wave can be more than six miles long.

solar cells—energy emitting cells that are powered by the sun.

spectrum—the complete range of a phenomenon. The spectrum of light includes all the colors of light. The spectrum of electromagnetic radiation includes all the wavelengths of electromagnetic energy.

spontaneous emission—the energy that is emitted from an excited atom as its electrons jump back to lower orbits.

stimulated emission—a special atomic condition needed to create laser light. In this condition, a photon is not absorbed when it strikes an electron but instead continues on its way. The electron in turn drops to a lower orbit and emits a photon. Thus two photons are created by stimulated emission. See **absorption, photons, electrons,** and **spontaneous emission.**

supersonic—faster than the speed of light.

thermonuclear reaction—takes place when the nucleus of an atom fuses under millions of degrees of heat. This fusion releases tremendous amounts of energy. The enormous power of a hydrogen bomb results from a thermonuclear reaction.

three dimensional—having the dimension of depth, along with height and width.

titanium—an extremely hard, light metal.

ultraviolet rays—very short light waves that cannot be seen.

vaporize—to make into a gaseous state.

vibrate—to move back and forth or up and down quickly and repetitively.

wave-front—the line or surface of light waves in which all its points are of the same phase. See **phase**.

wavelength—the distance between one wave crest and the next one that follows it. See **phase**, and **amplitude**.

Thomas Young—an English philosopher, physicist, and mathematician who devised important experiments with light (1773–1829).

H.G. Wells—an English historian and novelist (1866–1946).

BIBLIOGRAPHY

Adler, Irving. *The Story of Light.* New York: Harvey House, 1971.

Bixby, William. *Waves: Pathways of Energy.* New York: McKay, 1963.

Bronowski, Jacob, Jr., and Millicent E. Selsen. *Biography of an Atom.* New York: Harper & Row, 1963.

Burroughs, William. *Understanding Science: Lasers.* New York: Warwick Press, 1982.

Freeman, Ira Maximillian. *Light and Radiation.* New York: Random House, 1969.

Klein, Arthur. *Holograph.* New York: Lippincott, 1970.

――――. *Masers and Lasers.* New York: Lippincott, 1963.

Kohn, Bernice. *Light.* New York: Coward-McCann, 1965.

Lieberg, Owen S. *Wonders of Heat and Light.* New York: Dodd, Mead, 1966.

Mueller, Conrad. *Light and Vision.* New York: Time, Inc., 1966.

Nehrich, Richard B., Jr., et al. *Atomic Light: Lasers—What They Are, How They Work.* New York: Sterling, 1967.

Oldfield, Ruth L. *People of Destiny: Albert Einstein.* Chicago: Children's Press, 1968.

Stambler, Irwin. *Revolution in Light: Lasers and Holography.* New York: Doubleday, 1972.

――――. *The World of Micro-Electronics.* New York: Norton, 1969.

Ubell, Earl. *The World of Candle and Color.* New York: Atheneum, 1969.

Waller, Leslie. *Light.* New York: Grosset & Dunlap, 1968.

Weart, Spencer R. *Light: A Key to the Universe.* New York: Coward-McCann, 1968.

INDEX

Absorption, 20, 21
Airplanes, laser-propelled, 84
Al-Hazen, 12
Amplitude, 63–64
Argon laser, 8, 36, 77
Astronauts, 60
Atomic fusion, 78–83
Atoms, 19–20, 22, 58
  excited state, 25–26

Bell, Alexander Graham, 45, 46, 48
Bell Lab laser, 56–59
Birthmarks, removing, 37
Bohr, Niels, 20
Brain surgery, 37

Carbon dioxide lasers, 31, 51, 76
  computerized, 55
  cost of, 38
  medical uses of, 35–37
Cells and genes, 83–84
Chemical lasers, 32
Coaxial cables, stripping plastic off, 54–55
Cold lasers, 37, 84
Communications, 45–48, 70, 83
  with other star systems, 83
Computer circuit boards, 53–54
Cutting materials with lasers, 52, 60
Cylindrical hologram, 67

Deuterium, 81
Double-exposed holograms, 67
Drift, earth's, 60
Drilling, 9–10, 31, 51–52, 53, 55, 59
Dye laser, 59

Earthquakes, detecting, 60
Einstein, Albert, 16, 17, 19–20, 21–22
Electromagnetic energy, 25
Electrons, 20, 21, 22, 24, 26, 58
Ether, 16

Food packaging, 70
Frequency, wave-length, 25
Fusion fuel pellets, 82
Fusion reactions, 78–83

Gamma rays, 25, 44
Gas lasers, 31
Genetic engineering, 83–84
Gillettes, 49
Goldfinger (movie), 69
Greeks (ancient), 11–12
Ground state, 20

Helium neon gas lasers, 31
Hindus (ancient), 11–12
Holograms, 61–68
  amplitude and phase, 63–65
  interference, 66
  limitations of, 67–68

  meaning of, 61
Huygens, Christian, 13
Hydrogen atoms, 20
Hydrogen bomb, fusion reaction in, 79, 81

Industry, uses of lasers, 49–55
Information storage, 71–72
Ink, removing from paper, 55
Interference, 14–15, 66

Laser beams, 45–48
Laser fusion, 81
Laser light, 19–23, 27
Lasers
  acronym, 24
  in art and entertainment, 73–74
  communications, 45–48, 70, 83
  construction, 24–27
  future uses, 78–85
  glossary of terms, 86–92
  holograms, 61–68
  introduction, 9–10
  meaning of, 9
  measurement by, 56–60, 70
  medical uses of, 33–38, 70
  military uses of, 39–44
  peaceful uses of, 69–72
  safety standards, 75–77
  types of, 28–32
Lenard, Philipp, 16

94

Light, 11–18, 24
  Einstein's theory of, 16, 19–20
  interference, 14–15
  Plato's view of, 12
  Young's experiment, 13, 14, 16
Light waves, 15–16, 18, 47
  amplitude and phase, 63–65
  radiation, 24–25
Liquid lasers, 30

Magnetic confinement, 81
Maiman, Theodore, 26–27, 29
Mars, 22, 83
Maxwell, James, 16
Measurement, 56–60, 70
  chemical reactions, 59
  light pulses, 57
  smokestack emissions, 58
Microwaves, 26
Mode-locking, 29

NASA, 71
Newton, Isaac, 13
Nuclear power plants, 44
Nucleus, 20

Optical disks, 72
Optically pumped, solid lasers, 29, 33–34

Particle theory, 13
Phase, light wave, 64, 65
Photons, 16, 17, 21, 22, 24, 27
  speed of, 40
Photophone, 45, 46, 47
Planck, Max, 16
Plato, 12
Population inversion, 25–26
Printing, 9, 70–71
Pulsed lasers, 51, 55

Q switch, 29

Radar, 60
Radiation, 24–25
Radio waves, 25
Range finders, 40, 41
Retinas, laser repair of, 34–35, 36
Rockets, laser-propelled, 84
Ruby laser, 26, 27, 33–34, 50, 55

Safety standards, 75–77
San Andreas fault, 60
Sculptures, 74
Semiconductor lasers, 32
Smart bombs, 41
Smokestack emissions, 58
Space photographs, 71
Space weapons and research, 40, 41–44

Spontaneous emission, 21, 25
Steel, vaporized by lasers, 50
Stimulated emission, 22, 23
Stress testing, 68
Surveying, 60

Tattoos, removing, 37
The Who (rock group), 73
Titanium, 52

Ulcers, healing, 37
Universal Product Code label, 70

Video-disk systems, 72
Vietnam War, 41

Watch jewels, drilling holes in, 55
Welding materials together, 53
Wells, H. G., 39
Wounds, healing, 37

X-rays, 25, 44

Young, Thomas, 13, 14, 16

ABOUT THE AUTHOR

Brent Filson has published ten young people's books and more than 100 magazine articles, many of which deal with medical and scientific subjects. He currently lives in Massachusetts with his wife and four children.